The Battle for Oil

The Economics and Politics of International Corporate Conflict over Petroleum, 1860-1930

INDUSTRIAL DEVELOPMENT AND THE SOCIAL FABRIC, Volume 12

Editor: John P. McKay, *Department of History, University of Illinois*

INDUSTRIAL DEVELOPMENT AND THE SOCIAL FABRIC

An International Series of Historical Monographs

Edited by **John P. McKay**
Department of History
University of Illinois

Volume 1. **Forums of Order: The Federal Courts and Business in American History**
Tony Allan Freyer, *University of Arkansas*

Volume 2. **Competition and Regulation: The Development of Oligopoly in the Meat Packing Industry**
Mary Yeager, *University of California, Los Angeles*

Volume 3. **Keeping the Corporate Image: Public Relations and Business, 1900-1950**
Richard S. Tedlow, *Harvard University*

Volume 4. **The Growth of a Refining Region**
Joseph A. Pratt, *University of California, Berkeley*

Volume 5. **Business and Government in the Oil Industry: A Case Study of Sun Oil, 1876-1945**
August W. Giebelhaus, *Georgia Institue of Technology*

Volume 6. **Pioneering a Modern Small Business: Wakefield Seafoods and the Alaskan Frontier**
Mansel G. Blackford, *Ohio State University*

Volume 7. **The End of the Practical Man: Higher Education and the Institutionalization of Entrepreneurial Performance in Germany, France, and Great Britain — 1880-1940**
Robert R. Locke, *European Institute for Advanced Study in Management*

Volume 8. **Filling Up America: An Economic-Demographic Model of Popular Growth and Distribution in the Nineteenth-Century United States**
Morton Owen Schapiro, *Williams College*

Volume 9. **The Hungarian Economy in the 1980s: Reforming the System and Adjusting to External Shocks**
Josef C. Brada, *Arizona State University and* Istvan Dobozi, *Hungarian Academy of Science*

Volume 10. **The Management of Labor: The British and French Iron and Steel Industries, 1860-1918**
Judith Eisenberg Vichniac, *Harvard University*

Volume 11. **The Mitsubishi: Its Challenge and Strategy**
Yasuo Mishima, *Konan University*

Volume 12. **The Battle for Oil: The Economics and Politics of International Corporate Conflict Over Petroleum, 1860-1930**
A.A. Fursenko, *Translated and edited by* Gregory L. Freeze, *Brandeis University*

The Battle for Oil

The Economics and Politics of International Corporate Conflict over Petroleum, 1860-1930

by A.A. Fursenko

Translated and Edited by
Gregory L. Freeze

JAI PRESS INC.

Greenwich, Connecticut *London, England*

Library of Congress Cataloging-in-Publication Data

Fursenko, A.A.
[Neftianye voiny. English]
The battle for oil : the economics and politics of international corporate conflict over petroleum, 1860-1930 / by A.A. Fursenko ; translated and edited by Gregory L. Freeze.
p. cm.—(Industrial development and the social fabric ; v. 12)
Translation of: Neftianye voiny.
Includes bibliographical references and index.
ISBN 1-55938-262-7
1. Petroleum industry and trade—History—19th century.
2. Petroleum industry and trade—History—20th century.
3. Competition, International—History—19th century.
4. Competition, International—History—20th century. I. Freeze, Gregory L., 1945- II. Series.
HD9560.5.F8213 1990
338.2′7282′09034—dc20 90-20881

55 Old Post Road, No. 2
Greenwich, Connecticut 06836

JAI PRESS LTD.
118 Pentonville Road
London N1 9JN
England

ISBN NUMBER: 1-55938-262-7

Library of Congress Catalog Card Number: 90-20881

Manufactured in the United States of America

CONTENTS

Author's Preface to the American Edition

I was born in Leningrad and, after graduating from Leningrad State University, began to work at the Leningrad Section of the Institute of History of the Academy Sciences of the USSR. At one time here in St. Petersburg, the former capital of Tsarist Russia, were located the chancelleries of ministries and the offices of business enterprises. The papers of these are now preserved in the Central State Historical Archive, located in one of the largest building complexes in the city—the former residences of the Senate and Holy Synod. Documents from this archive constitute the primary core of materials used in the preparation of this volume.

Shortly after I began working at the Institute of History at the end of the 1950s, a group of scholars from the Academy of Sciences and staff members of archives commenced preparation of a two-volume publication entitled "Monopoly Capital in the Russian Petroleum Industry," embracing the period from the 1880s to 1918. That project laid the foundation of my subsequent research on the oil problem and led to a number of articles and books on the subject.

But I should add that two individuals also played a major role in stimulating my interest in the history of Russian oil and making this subject a primary focus of my scholarly research. One major influence was my father, Aleksandr Vasil'evich Fursenko, a trained geologist and specialist on petroleum who devoted his entire life to explorations for oil in the USSR and who also established a scientific school in the field of micropaleontology.

He would often recall the words of the famous chemist, D.I. Mendeleev (who for many years had served as a consultant at the tsarist Ministry of Finance) that using oil to fuel locomotives and steamships was like burning money in a stove. My father travelled around various oil-producing areas of the Soviet Union and told how, after the Revolution of 1917, documents of the former oil companies were used to wrap herring. He observed that it would be interesting to try and reconstruct the history of the oil business by using those documents which remained intact.

The second person, to whom I am indebted for my professional training as an historian, was my teacher—a distinguished scholar and wonderful human being, Boris Aleksandrovich Romanov. Romanov's life was anything but easy: at the end of the 1920s he was arrested (along with several other leading historians and scholars) in the so-called "Academy Case" and spent more than a year in prison, and then another ten years in exile. However, prior to his arrest, Boris Aleksandrovich managed to work in what was then called Glavarkhiv. He was among the first to obtain access to documents from the former tsarist chancelleries and published several interesting works on the economic and diplomatic history as well as on the history of the working class in Russia at the end of the nineteenth and beginning of the twentieth centuries. When a severe flood inundated Leningrad in 1924, Boris Aleksandrovich was one of those who took an active role in rescuing documents that had been submerged under water and then seeing to their drying and preservation. Many of the documents that he saved (in particular, the materials of the St. Petersburg International Commercial Bank) were used in my book and in other publications. In the 1940s Boris Aleksandrovich returned to teach at Leningrad State University, where I became acquainted with him and became one of his students and followers. He was one of those who supported my work in the archives for the documentary collection on the oil business.

Although the history of oil has been the primary focus of my research, I have also worked on a variety of other subjects during my tenure at the Institute of History. One major focus has been the history of the United States from the American Revolution to the present; in particular, I have given special attention to foreign policy, particularly with respect to the origin of the "open-door policy" in China. For that I used a large corpus of documents that had been assembled on this question by the tsarist Ministry of Finance, which was then headed by S. Iu. Witte—a virtual minister of foreign affairs in matters affecting the Far East. Witte's ministry played a major role in determining Russian economic policy in the late nineteenth and early twentieth centuries; hence that ministry's archives are also exceedingly important for a study of the oil question. Various aspects of the history of the oil business and policy became one of the central themes during the many years I have worked on this subject.

The Academy of Sciences has repeatedly sponsored research trips to the United States, Great Britain, France, and Rumania. The present volume draws heavily upon the documents that I found on the oil question in the archives of those countries. I would like to take this opportunity to express my gratitude to the École pratique des hautes études in France, whose director, the late F. Braudel, invited me to come to Paris for research and who helped arrange my access to materials in the archive of the Ministry of Foreign Affairs. In May 1985 I was able to spend only a few days in the Public Record Office in London, but thanks to its splendid and efficient organization and to the assistance of my friend, Dr. Ronald Ferrier, I was able to accomplish a great deal of intensive research. In 1962 I also had the opportunity to work for one month in the archives and libraries of Bucharest and Ploiesti in Rumania, where I was given valuable assistance by the Institute of History of the Rumanian Academy of Sciences.

I would especially like to note the role in my research that has been played by the chance to work in American archives where, altogether, I have been able to work several months. Most important, I have been able to work in the Rockefeller family archives; I first conducted research there in September 1968 and proved to be the first foreign scholar to have the privilege of using these materials. I would like to use this occasion to express my profound thanks to the family archivist, Dr. Joseph Ernst, a brilliant expert on the Rockefeller documents who laid the foundations to the organization of the Rockefeller Archive Center. Since his retirement he has been succeeded by a skilled new director, Dr. D. Stapleton. As for my research on manuscript materials at the Library of Congress and the library of Princeton University, I should also like to express gratitude to my friend, James H. Billington, formerly a professor at Princeton and presently the Director of the Library of Congress. In 1976 I also enjoyed his hospitality as a visting scholar at the Wilson Center in Washington, where he was then the director. In the National Archives of the United States I was assisted by Karl Weissenbach and the head of the sector for the preservation of documents on the army and navy, Vanderhoff.

I also would like to express my thanks to the International Research and Exchanges Board (IREX) and, especially, to its administrators, Allen Kassof and Daniel Matuszewski, who were my sponsors during my scholarly research trips to the United States and gave them their unfailing support. I am grateful too to Dr. Gregory Guroff, who also rendered invaluable assistance at various points during the many years of my research.

The first American historian whom I met (in 1958) was Professor Theodore von Laue, author of the book *Sergei Witte and the Industrialization of Russia*. For over thirty years we have remained in cordial contact, exchanging opinions and literature. I would also like to acknowledge my friends Ruth and Robert Roosa, with whom I am bound by more than a decade and a half of collaboration and coopration. I have participated with Ruth Roosa, a specialist in Russian economic history, in scholarly conferences and joint publications. I am most grateful to the Roosas both for their assistance in obtaining access to documentary materials (indispensable in the preparation of this volume) and for their friendship and support of my research.

Ever since I began working at the Leningrad Section of the Institute of History, I have enjoyed the support of my colleagues for all these many years. But with respect to this book, I am particularly indebted to my friends and colleagues, R. Sh. Ganelin and B.V. Anan'ich—also students of Boris Aleksandrovich Romanov. B.V. Anan'ich was also the official editor of the original Russian text for this volume. Preparation of the American edition was assisted by my students at the institute, V.N. Pleshkov, S.A. Isaev as well as S.K. Lebedev—a student of B.V. Anan'ich.

The idea of publishing this volume in the United States emanated from a well-known scholar and specialist in the domain of Russian economic history, Professor John McKay, who published an extremely generous review of this work in the *American Historical Review* (1988). I am grateful to him for the kind review of my work and for his assistance in arranging the publication of my book in English. I am also much indebted to Professor Gregory Freeze, who is well-known in the Soviet Union as a productive scholar in the much-neglected domain of Russian religious and social history. Endowed with a phenomenal capacity for work, he has impressed the staffs of archives and libraries as well as his scholarly colleagues with his devotion to scholarship. I am very grateful to Professor Freeze for his labors on my behalf.

A special word of gratitude is also due to my publishers. I am deeply indebted to the Leningrad branch of "Nauka" press, which published the original Russian edition of *Neftianye voiny* as well as all my other books and which has also taken an active role in arranging their publication. I am also grateful to JAI Press for its willingness to assume the burden of making the volume available to the English-speaking world.

Finally, I would like to say that, at every stage in the preparation of this volume, I have enjoyed the unstinting support and assistance of my wife, Natasha. Without her help this book could not have been written.

A. Fursenko
17 March 1989
Leningrad

Editor's Note

This volume is a much revised and expanded version of A.A. Fursenko's Russian text, originally published as *Neftianye voiny (konets XIX—nachalo XXV)* (Leningrad: Izdatel'stvo "Nauka," 1985). The author has substantially reduced the sections on American oil trusts (based largely upon Western scholarship) and greatly expanded those sections that draw upon his own archival research, particularly for materials pertinent to the Russian case. Thus each chapter has undergone substantial revision and one entire new chapter (seven) has been added.

So far as the translation itself is concerned, I have tried to render the text as freely but as precisely as possible. To avoid an erroneous double translation of quotations that were not originally in Russian, the author has supplied the original text, which is either given directly (if in English, exactly as recorded by the author) or translated directly into English (from French and German).

Transliteration follows the conventions of American Slavic scholarship. In the case of well-known personalities (such as Witte or Nicholas II), the spelling conforms to that conventionally used; the same rule applies to figures (such as Mantashoff) who were well-known in the contemporary West and have entered the monographic literature under that spelling. For the spelling of lesser known Western figures, the author has supplied the form that he encountered in the documentation. Wherever appropriate, dates are given in both the prerevolutionary

Russian calendar and the modern Gregorian calendar; the former (based on the Julian calendar in use in Russia until 1918) lagged twelve days behind the Western calendar in the nineteenth century and thirteen days in the twentieth century.

—*Gregory L. Freeze*

Introduction

Almost an entire century has passed since the oil problem became inextricably connected with international relations. At the beginning of the twentieth century V.I. Lenin noted that the oil monopoly plays a key role in politics, that its actions are an "instructive example" of the struggle to "divide up the world" among the alliances of capitalists that he called characteristic of the most recent stage of capitalism.[1] World production of oil by capitalist countries amounted to 20 million tons in 1900, rose to 50 million tons by 1913, but by 1980 had mushroomed to 2,354 *billion* tons. Whereas petroleum provided only 5 percent of the energy component for capitalist countries in 1913, this had grown to 45 percent by 1980.[2]

The mounting importance of alternative sources of energy and vigorous efforts at energy conservation resulted in a reduction of petroleum consumption to 1.920 billion tons by 1985. Oil is of course one of the irreplaceable sources of energy; based on energy consumption levels in 1985, known reserves only suffice for the next 45 years. In 1985 the United States imported 152 million tons, but that represented a fifty percent reduction in import levels since 1977. The total cost of oil consumption by all branches of the American economy represented 155 billion dollars. Within this consumption, each year an increasing share of oil goes to the petro-chemical industry, whose volume of production reached 80 billion dollars in 1985.[3] In the future the primary usage of oil will lie precisely in the chemical industry, although today it still constitutes an important source of fuel and energy.

Compared with the beginning of the century, the assets of oil corporations have grown from tens of millions of dollars to tens of billions of dollars by the 1980s. In early 1985 it became a sensation when the television company ABC acquired an enterprise from the entertainment industry (Capital) for 3.5 billion dollars—an acquisition rated as the largest merger of capital in American history. But this deal pales in significance when compared with those in petroleum. Thus in 1984 one of the "seven sisters" of the international oil cartel, the American firm "Standard Oil of California" (Chevron), devoured one of the other American sisters (Gulf Oil Corporation), in a deal involving 11 billion dollars.

In 1929, when a list of the twenty five largest industrial corporations of the United States was first compiled, eight of these were petroleum firms. These included Standard Oil of New Jersey (Exxon), which was second on the entire list, and Standard Oil of New York (Mobil), listed in seventh place. A half century later many of the corporations on this list have disappeared or dropped to the second tier, but the petroleum firms have significantly strengthened their position and surged upward: Exxon has risen to first place, Mobil to third place. As a result, by 1979 thirteen of the twenty five largest firms were in oil.[4] Within the scale of world capitalist economy, the International Oil Cartel continues to operate: first formed in 1928 and composed of seven firms until 1984, it now has six "sisters"—four American, one British, and one Anglo-Dutch corporation. These are all among the largest trans-Atlantic corporations.

Beginning in the late nineteenth century, the role of the oil factor has steadily risen in the economic and political life of countries throughout the world. It is utterly impossible to understand either the origin of the world wars or the underlying causes of many other international conflicts without taking into account the oil problem.

The beginning of the fight for oil goes back to the appearance of the first oil trusts—monopolistic associations which sprang up alternately in America, Russia, Great Britain, the Netherlands, and other countries. Petroleum corporations had already attained gigantic dimensions, great economic power and political influence. They pursued their own policy in domestic politics and became actively involved in international affairs. The battle for oil became an organic component of the world imperialist conflict over the struggle for sources of raw materials, markets, monopolistic profits, and spheres of influence. These comprised an integral part of international relations.

The character of the oil wars evolved in accordance with the changing significance of oil—first as a product for purposes of illumination (kerosene), then as a source of energy for internal combustion engines (fuel oil, diesel oil, and gasoline), and lubricating oil for machinery and equipment. The competition for oil passed through different stages, which were linked to changes in the sphere of competition: at first this was a trade war for markets, then a struggle for control over oil resources, and finally a bitter fight for oil deliveries that has become a vitally important factor since the onset of the energy crisis in the 1970s.

Numerous books and articles, of the most diverse genre, have focused on the oil problem and shed light on various aspects and various historical periods. This problem is multi-faceted and virtually unlimited, both in terms of its subject and the volume of material available to the researcher. The importance that this question has recently acquired as one of the most important factors in the world economy and international relations has increased interest in the oil crisis and its historical background. These considerations have shaped this study, which seeks to describe the battle for oil in the late nineteenth and early twentieth centuries—events which laid the basis for the developments that have since become so critical and so dangerous for the world.

About two-thirds of the presently known oil reserves of the capitalist countries are now found in the Near East, including Saudi Arabia, Kuwait, the United Arab Emirates, Iran, and Iraq. It is from this gigantic reservoir that oil streams to America, Western Europe, and Japan. But the Near Eastern deposits were opened only shortly before World War I, and their intensive exploitation began still later. Many years before this the main producers of oil and its main suppliers on the world market were Russia and the United States. In the United States the Rockefeller oil trust of "Standard Oil Company" was formed, while in Russia the Nobel and Rothschild oil corporations, likewise had a character similar to that of a trust. The conflict between the Rockefeller trust and Russian oilmen laid the basis for the first battles over oil on the international area.

A bitter competition on the oil market was accompanied by a fierce battle on domestic markets to establish complete monopolistic control. Hence the oil corporations had to conduct their war on two fronts, and indeed they would have to do this for several decades. At the initial stage, the domestic front had particular importance, for a solid rear base was an essential precondition for success on the international front. But subsequently the foreign operations served as an important means of strengthening the monopoly at home as well. To this it should be added that the character and degree of monopolization in various countries was shaped by the conditions of their historical development, which exerted a direct influence on the form of activity of oil corporations.

NOTES AND REFERENCES

1. V.I. Lenin, *Polnoe sobranie sochinenii* [hereafter *PSS*] (Moscow, 1958-1965), 27, pp. 367-368.
2. *United Nations Statistical Yearbook* (New York, 1980); *British Petroleum Statistical Review of World Energy: 1986* (London, 1986).
3. *Time,* April 14, 1986.
4. *Fortune,* May 5, 1980; A.E. Primakov, *Persidskii zaliv: neft' i monopolii* (Moscow, 1983), pp. 123-124.

Chapter I

The Beginnings of International Oil Competition

THE FORMATION OF ROCKEFELLER'S AMERICAN TRUST

The roots of the American oil trust, The Standard Oil Company, go back to the mid-1860s. After Edwin Drake discovered oil deposits in Pennsylvania in 1859, a great number of companies were soon formed to explore, extract, and refine oil and also to organize the marketing of oil products. In Pennsylvania and the neighboring state of Ohio these oil companies sprang up like mushrooms. Their main product was kerosene, which stimulated substantial demand for use as a means of illumination. The kerosene business brought in large profits, and that inspired John D. Rockefeller to form the Standard Oil Co., which subsequently grew into a gigantic corporation. Cleveland, Ohio—Rockefeller's residence—became one of the centers of a full-scale oil fever.

The capital that Rockefeller invested in oil had been amassed from his business as a supplier for the American army during the Civil War in 1861-1863. Not only in America, but in many other countries as well, much capital derived originally from business with the military. Rockefeller accumulated his wealth by supplying agricultural products and other provisions, an activity that yielded phenomenal profits under wartime conditions.

By the time the oil boom commenced, the opportunities for profit in dealings with the army had been exhausted. Rockefeller made his very first investment ($4,000) in an oil enterprise in 1863, two years later he already owned a plant to produce kerosene, and after another two years he had five such plants in his company's possession. By 1870 the company had a capital of one million dollars and ranked as the largest oil enterprise in America.

The American magazine *Fortune* has written that "it would be hard to choose a large company that illustrates the evolution of U. S. capitalism better than [Rockefeller's] Standard Oil Company (N.J.)."[1] That statement is apparently correct, at least in the sense that the history of this oil trust's emergence and growth provides an instructive example of the patterns of capitalist development in the United States. Endowed with extraordinary business skills, Rockefeller created a powerful company where production attained a high degree of organizational and economic development. The company was carefully structured and organized; this was an important advantage, bestowing great economic effectiveness and enabling Rockefeller to overcome less successful competitors. From time to time Rockefeller cut prices in order to bankrupt rivals and then buy up their property. Price wars became his chief weapon for eliminating competitors.

To be sure, the Standard Oil plants also made technological innovations part of the formula for success. But the key factor in the rise of the Rockefeller company was its speculative activities. And in this respect, Standard Oil was quite distinct from American firms like the Ford automobile company or the Edison electrical company, which, to a significant degree, owed their success to achievements in engineering. Ford and Edison were outstanding designers and inventors, not merely great entrepreneurs. But Rockefeller was a master at financial dealings and speculative operations, a specialist without peers in the game of behind-the-scenes machinations.

"History is more or less bunk," said the automobile king Ford, hoping thereby to emphasize that there is no point in recalling the past or looking back on yesterday.[2] That is, one must live for today and tomorrow. Why look back? For his part, Rockefeller never had any doubts. He has been accused of abuses, and some published materials have clearly documented his company's illegal activities. With unconcealed animus, Rockefeller castigated those who published these materials and in his private correspondence he always put the words "history" and "historian" in quotation marks. Was there any point in looking back into history if Standard

Oil succeeded in eluding judicial prosecution and criminal punishment? "They wrote 'histories,'" Rockefeller later commented, "but the Standard Oil Company wrote history in its doings, which will stand through all time."[3]

Like medieval knights, Rockefeller's profits made him a robber-baron. But that did not disturb him in the slightest. Many years later, as an old man travelling through France, he said the following as he sat alongside a medieval castle: "The robber-barons of those days had a hard time.... They had no attorney generals to protect them, so they had to build fortresses and shut themselves in."[4]

Rockefeller did not make his affairs public, but, on the contrary, attempted to keep them secret. If something did leak out, he took steps to show Standard Oil as filled by the most innocent intentions. That is how he behaved, even in those instances when the most flagrant violations of the law had come to light. A good case to the point is his secret deal with railway companies to give Standard Oil preferential conditions and tariffs over her rivals.

The problem of transporting petroleum products had always been fraught with great significance, since the freight charges played a major role in determining the competitiveness of oil firms. Hence Rockefeller always attached great significance to plans for a deal with railway companies. The American writer J. Abels notes that, in Standard Oil's growth and rise to predominance, preferential railway tariffs played a more important role than the managerial skills of Rockefeller and his partners.[5]

The task of reaching an agreement with the railroads was assigned to a Standard Oil partner, J. Flagler, endowed with a special ability in negotiating. "The ability to deal with people is as purchasable a commodity as sugar or coffee," Rockefeller once said, "and I pay more for that ability than for any other under the sun.[6] He had close ties to Flagler; they were not merely partners, but people of a like mind. They spent their workday together in the same office building, lived near each other on the same street and travelled to and from work together, discussing business matters along the way. They attended the same church and often spent their evenings together. "It was a friendship," as Rockefeller later wrote in his memoirs, "founded on business, which, Mr. Flagler used to say, was a good deal better than a business founded on friendship, and my experience leads me to agree with him."[7]

Beginning in 1867 and for the next three years, Flagler more than once made agreements that gave special rebates in exchange for an obligation to charter a substantial number of tank cars for hauling kerosene. The railroad was interested in securing a regular supply of cargoes and utilization of rolling stock and in eliminating periods when the cars stood idle. Hence railroad companies were eager to make agreements with Rockefeller, whom they regarded as a reliable business partner. To strengthen its reputation, Standard Oil chartered empty tank cars beyond the number fixed by the agreement, and in return obtained rebates for the transport of its kerosene.

In a short time, Rockefeller practically monopolized the oil refining business in Cleveland, absorbing 21 of 26 local firms into his own enterprise. At the same time, Standard Oil purchased a kerosene plant and commercial firm in Long Island, thereby strengthening his position in New York. As a result of the oil war in 1872, Rockefeller gained control over 25 percent of the oil refining enterprises in the United States. He held out an olive branch to his most dangerous rivals in other states, promising them a position in Standard Oil if they would join it. The companies of W. Warden and C. Lockhart began to buy up all the oil refining companies in Philadelphia and Pittsburgh, with the aim of merging them with Standard Oil. Analogous agreements were concluded with the firm of C. Pratt and C. Rodgers of New York, J. Vandergrift of Titusville, and also with the Pennsylvania company of John Archbold (who would later become Rockefeller's right-hand man). According to the formal agreement, the new partners of Standard Oil were nominally independent and, as they bought up other enterprises, pretended to be fighting Rockefeller. In response to rumors that these companies had "sold out to Standard Oil," Rodgers, Archbold and other participants in the deal issued public denials.[8] This entire operation of making agreements with former rivals was completed by 1875, when Standard Oil's capital had increased to 3.5 million dollars. Just three years later, Standard Oil already controlled 90 percent of the total capital investment in oil refining in the United States.

In the beginning, Standard Oil did not engage directly in oil production, preferring instead to concentrate on refining and marketing the oil products. The reason is that oil production was the least profitable, most risky phase of the process. That is evident from the company's profits in 1891-1911, when Rockefeller already possessed a secure position in oil production. Of the company's total profit (1,280 million dollars), it "earned" 532 million on transport, 307 million from trade, 259 million from refining, and only 170 million from the production of crude oil.[9]

There were good reasons for Rockefeller's decision to concentrate upon the refining and marketing phases of the industry. The production of crude oil required enormous capital investments, but those resources only became available at a later stage in the development of Standard Oil, when the company had a stronger financial base. It was also a high-risk phase, since only a small percentage of prospecting and drilling operations actually proved economically feasible. At the same time, monopolization of oil refining and marketing was a lever enabling Rockefeller to establish his power over the oil producers. That is why the monopolization of refining and marketing preceded the involvement in oil production. From the very beginning Rockefeller strove to secure Standard Oil's complete dominance in the American petroleum industry. Whereas Rockefeller controlled 80 percent of the 10 million barrels of oil produced in 1876, his share had increased to 95 percent of the more than 46 million barrels produced a decade later.[10]

By the early 1880s Standard Oil had grown into a gigantic national enterprise, and the company's management now faced the question of redesigning a more integrated structure. Several companies, under the control of Rockefeller and his partners, were formally combined in 1882. According to a secret agreement, the owners of these companies (which engaged in oil production, refining, transportation, and marketing) gave their stocks to nine trustees: John and William Rockefeller, O. Payne, C. Pratt, J. Flagler, J. Archbold, W. Warden, J. Bostwick, and B. Brewster. In exchange for their stocks, the former stockholders received trust certificates, but in return surrendered any influence over the conglomerate's affairs.

The capital of the oil trust constituted an enormous sum—70 million dollars. Two thirds of the capital (46 million dollars) was in the hands of the "supreme council," where Rockefeller's influence was doubtlessly predominant.

The formation of the trust was accompanied by the creation of two new divisions in the northeastern states of New York and New

Jersey—the financial center of the country, a fact which predetermined a special role for them. For a considerable period of time, Standard Oil was self-financing and did not rely upon the services of a bank. It was only in the late 1890s that the company called upon the services of the National City Bank of New York, whose president (James Stillman) later became tied by kinship with Rockefeller's family: two of Stillman's sons married the daughters of William Rockefeller, the brother and closest partner of John Rockefeller. The National City Bank joined the system of the Rockefeller oil trust, even earning the accolade of "the Bank of Standard Oil." But until then, the trust's financial transactions were handled by Standard Oil of New York, the central branch of the conglomerate for over 15 years.

New York was the capital of financial and industrial life in the United States. The tiny Wall Street in southern Manhattan, together with contiguous blocks, was jammed with the offices of the largest banks and industrial companies. On May 1, 1885 Rockefeller's company moved its headquarters to a private building at 26 Broadway (perpendicular to Wall Street). Although formally registered as the property of the Standard Oil Company, this was the headquarters for the whole trust. That is how things remained even when antitrust legislation forced a transfer of the main central organizational functions to Standard Oil in New Jersey (because this state offered the most liberal incorporation laws, which enabled evasion of federal laws with impunity). Although since that time (1899) the chief company of the oil trust became Standard Oil Co. of New Jersey, its real headquarters, as before, remained in New York—which, for business purposes, was far more convenient.

Formation of the trust and relocation of its central offices to New York opened a new epoch in the activity of the Rockefeller company. The company thus took steps to expand its oil refining enterprises. Between 1882 and 1885 their total number declined from 53 to 22, but the three largest plants (located in Cleveland, Philadelphia and Bayonne, New Jersey) produced over 40 percent of all distillates.[11] Simultaneously, the company sought to control the domestic market by establishing control over the retail and wholesale distribution of kerosene in the largest cities of America. The landscapes of New York, Cincinnati, Chicago, and other cities were now reshaped by the enormous storage tanks of Standard Oil, which held a virtual monopoly over the sale of kerosene.

Rockefeller's trust wielded enormous economic power and used its influence to secure the protection of local and federal authorities. Samuel Dodd, who for many years served as the legal consultant for Standard Oil, asserted in his memoirs that "I know of no instance of the Standard Oil Company desiring legislation in its own favor." To be sure, admits Dodd, as the company lawyer he "often opposed" various pieces of draft legislation, "but so far as I know," Standard Oil always "used legitimate means in its opposition."[12]

The term "legitimate means" connoted whatever would guarantee Rockefeller's interests—a philosophy based on the sacred principle that money is power. One of the greatest magnates of the day, Collis Huntington, urged financiers to become actively involved in state politics. "If you have to pay money to have the right thing done, it is only just and fair to do it," said Huntington.[13] Rockefeller's public statements did not indulge in such candor, but behind the scenes he followed the very same policy.

Standard Oil, it is fair to say, systematized political graft as an instrument of company policy. Shortly after the turn of the century, secret documents from the chancellery of a Rockefeller partner (John Archbold) that became public revealed the bribery of politicians and their transformation into company agents. The veracity of the documents are beyond any doubt. It has been claimed that Rockefeller himself was not directly involved and indeed knew nothing about such things. Frankly, it strains all credulity to accept such assertions, since Archbold was Rockefeller's most trusted person. "He filled a peculiar position in our organization, which no other man did fill," as Rockefeller himself later admitted.[14]

Legislative organs and other higher instances, including the federal bureaucracies and the very pinnacle of executive political authority, were subject to pressure from Rockefeller. For example, the Dodd memoirs recount how Standard Oil relied upon the support of a group of people in the New Jersey state legislature who were identified as the "Black Horse Cavalry."[15] They helped to divert funds to state chancelleries; once the company had fed the Black Horse Cavalry, it had no need to worry about decisions in New Jersey politics.

At the end of the nineteenth century, American development entered a phase of corporate dominance. Hereafter it was not random attacks, but unrelenting, well-orchestrated pressure that put a qualitatively new level of pressure on the government and its policies. With the appearance of this new stage in the development of capitalism, the political profile of the country changed as well. As we shall see, this led to a reconstruction of the administrative system in the United States.

The Rockefeller oil trust became so dominant a force that it justifiably came to be called a "state within a state." Having established virtually undivided predominance within the United States, Standard Oil embarked on a struggle for the world market, but here it would encounter far more difficult battles.

Its first shipment of kerosene to the international market came in the early 1870s, but by the mid-1870s exports already constituted a third of the company's commercial activities. Initially not encountering any resistance, Rockefeller conquered the markets of Germany, England, France, Russia, India, China, and many other countries. According to the U.S. Geological Service, American kerosene "had penetrated the most distant parts of the world."[16] Its monopoly on foreign markets enabled Standard Oil to sell dearly and reap tremendous profits. These earnings, in turn, gave Rockefeller a great advantage on the domestic American market, for his company could more easily sell at reduced prices to ruin competitors. In other words, what Standard Oil lost in the United States it could recoup, with interest, in exports.

THE RUSSIAN OIL BUSINESS: THE NOBEL BROTHERS' COMPANY

In the early 1880s a serious rival to the American trust appeared on the world market—Russian oil companies. This competitor's oil production was at first concentrated in Azerbaidjan, on a small piece of the Apsheron Peninsula occupying some fifteen square kilometers. The capital city of the region, which had begun to produce oil in the 1870s, was Baku; it was here that the Nobel Brothers founded an oil company in 1879 that would later grow into an enormous corporation. By the end of the century the Nobel corporation had amassed enormous power, but as early as the mid-

1870s it already claimed a leading role in the Russian oil business, ranking first among firms engaged in oil producing and marketing.[17]

The joint-stock company that the Nobel Brothers established in 1879 possessed a capital of 3 million rubles. Its heads were two native-born Swedes, Ludwig Nobel and his brother Robert—brothers of the famous inventer of dynamite, Alfred Nobel. They had come to Russia with a small capital, but then successfully increased their wealth by furnishing weapons for the Russian army, especially during the Russo-Turkish War of 1877-78. The profits from these arms sales provided the core capital for the Nobel oil company. The Nobels leased their first oil tracts in 1874 and bought their first kerosene plant in 1875, but it was only in the early 1880s that their enterprise acquired national significance. By 1882 its capital had increased to 10 million rubles, and two years later swelled to 15 million rubles.[18]

Like Rockefeller in America, the Nobel firm began by concentrating on the secondary phases of the petroleum industry, that is, distillation, acquisition of a transportation network, and controlling the sale of petroleum products.[19] The company reconstructed and expanded its oil refining plant according to the most modern technology, which enabled it to obtain kerosene not only in larger quantities, but also with a higher quality. By the late 1870s the Nobel plant was already the largest in the country, and in 1885 the volume of its output exceeded the total production of the five largest rivals operating in Baku.

Since the primary means of transportation in postreform Russia was still waterways (the Caspian Sea and Volga River), by the late 1870s Nobel was actively studying and organizing the most economic means for shipping oil products by water. In 1878 the company launched its first large oil tanker, operating between Baku and Astrakhan; by 1885 the firm already owned seventeen tankers, two of which were delivering oil to the Baltic. A decade later the firm had one hundred tankers of various sizes to deliver 4 million barrels of kerosene annually to the port at Astrakhan. The sheer efficiency of operations, especially for that time, is impressive: the largest Nobel tankers in Baku could be pumped full of kerosene in three hours; they required 34 hours to reach Astrakhan by sea; then in Astrakhan they needed another 4 hours to disgorge their storage tanks. At the next stage, hundreds of smaller vessels delivered the kerosene to numerous points along the Volga, and from there the kerosene was hauled overland to the Russian interior.

The main staging post for distribution inland was Tsaritsyn, where the Nobel firm built gigantic storage tanks. For shipping by rail Nobel constructed tank cars, which were based upon American prototypes. In 1883 the firm had sixty railway units, each consisting of 25 tank cars, moving along the spurs of the Russian railway. In 1885 the firm had storage facilities in 40 rail centers; by 1900 it had increased this number to 129. The largest distribution centers were Nizhnii Novgorod, Rybinsk, Iaroslavl', Samara, Saratov, Kazan', Perm' and Tver'. The firm made its first delivery to St. Petersburg in 1881, and later constructed enormous oil storage tanks on the outskirts of the city. It did the same in Moscow, Riga, and Domino (a gigantic distribution center near Orel that was equipped with a thick network of access roads). By the end of the nineteenth century, the firm's rolling stock included thousands of tank cars.

For points outside the railway network, the company had to deliver kerosene in barrels using alternate forms of transport, such as horse-drawn carts. To facilitate distribution, the company built special plants to produce the requisite barrels.[20]

Along with kerosene, from the early 1880s the Nobel company became very active in marketing residual fuel oils—that is, the petroleum residues left after the refinement of kerosene. To stimulate demand for fuel oils, the Nobel machine-building plant in St. Petersburg succeeded in producing an improved fuel injector, which, with the help of a spray of steam, atomized the fuel oil and raised its inflammability. Such fuel injectors were installed under the boilers of steamships and locomotives as well as in various industrial enterprises, and the result was a sharp increase in demand for fuel oil, which thus became an important source of energy.

By the turn of the century, internal combustion engines also became more widely used. The Nobel plant concluded an agreement with the inventor, Diesel, to begin production of the new engines. Even as the Nobel company launched production of diesel engines, it did not seek to reduce production of fuel oils. The latter still brought in an excellent profit and guaranteed a strong demand for the refinery's residues, the profits from which steadily increased. By the end of the nineteenth century these constituted a ten-fold increase since 1883.[21]

The profitability of the Russian oil market encouraged a still more accelerated pace of exploration and drilling, causing in turn an increased importance of petroleum for the country's energy and fuel balance. According to calculations on the consumption of fuels for industry and transportation, oil achieved rough parity with coal during the period of 1888-1914.[22]

The sale of fuel oil pushed the kerosene business into second place, and by the early 1890s the Baku petroleum industry had become more occupied with fuel oil than kerosene. According to statistical data for 1893, the output of fuel was one and half times that of kerosene.[23] However, these figures do not provide a full picture of how things actually stood, for the kerosene trade yielded very high profits and a large portion of this output went for export. It was no accident that, in the annual reports of the Nobel Company, the income from kerosene sales significantly surpass the earnings from fuel oil sales.[24] The fuel oil business on the domestic market, however, was still significant, for it provided a secure rear base amidst the oil wars with the Rockefellers on the world market. From the mid-1880s this international oil competition significantly intensified, but it took the form of a "kerosene war" to control the markets and sale of kerosene.

At first, the Nobel (like Standard Oil) made relatively small investments for the exploration and production of crude oil. In 1881 such investments constituted only 5.3 percent of company assets, although three years later these had risen to 8.1 percent. At the same time, the investments in refining and transport had risen sharply. In 1885 the plant property represented 29.6 percent of company assets, whereas transportation facilities comprised 62.3 percent (including 34.3 percent in ships and barges, 19.5 percent in storage tanks, and 8.5 percent in oil pipelines).[25]

By the mid-1880s the Nobel concern held the leading position in the Russian oil business and controlled powerful financial resources. The firm's high profits were an important factor in the concentration of capital and the monopolization of industry and trade. As early as 1883 Nobel already controlled more than a fourth of all petroleum production in Russia and about half of the kerosene business. Although Nobel established ties with banks, it had all the characteristics of modern capitalist organizations (of the "trust" category) and, like Standard Oil, was essentially self-financing. It first accumulated its capital through government contracts, but obtained capital for subsequent investments from the exploitation of the Baku petroleum industry. Despite the fact that the corporation owners were Swedish by origin, the company was hardly a foreign firm. Although the firm, both at the outset and later, turned to foreign capitalists (Swedish, French, but above all Germans) for assistance, that did not violate its economic independence. The independent status of this company—the largest monopolistic corporation in the Russian petroleum industry—was also due to Nobel connections with tsarist elites and the support of authorities.[26]

This support took different forms, including financial assistance. In 1883 the State Bank gave the Nobel Company a credit for 2 million rubles at 7.5 percent interest—at a time when private banks were charging 17 to 18 percent interest.[27] In accordance with a decision by the Ministry of Finance, the State Bank regularly renewed this credit through the first decade after 1900, although the company requested such assistance only when absolutely necessary and paid its debt punctiliously. "This credit (opened in the form of a special current account)," reported the Council of the State Bank after its next review, "has been used properly—that is, the indebtedness is quite volatile and several times each year drops to zero." It added that "apart from this credit, E. L. Nobel and the joint-stock company 'Nobel Brothers' have commercial credits at the St. Petersburg Office of the State Bank (the former in the amount of 500,000 rubles, the latter for 1,500,000 rubles), but they have made almost no use of these credits."[28]

In extending support to Nobel, the government was seeking not only to promote the development of its enterprises, but also to extract advantage for itself, for the petroleum business was an important source of state revenues. The government established an excise on kerosene and thereby obtained millions of rubles in new tax revenues. In 1877, under the pressure of oil businessmen (above all Nobel), the kerosene excise was abolished, and for more than a decade the Baku industrialists remained free from this tax. That permitted them to invest additional means in the petroleum business. In 1888, however, when the Russian petroleum business had become a branch of production of enormous size, the excise tax was reintroduced. Although it applied only to kerosene sold on the domestic market (not to petroleum products being exported), the levy brought in enormous revenues to the state treasury:[29]

Year	*Excise Revenues (rubles)*
1888	6,607,780
1889	9,298,024
1890	10,567,743
1891	10,174,758

The government also earned a profit from the transport of petroleum products on the state railway lines, which set rather high tariffs. The development of kerosene exports abroad also created an opportunity to accumulate foreign currency.

The petroleum industry remained under the constant surveillance of state authorities. Its development was strictly regulated and subject to state supervision. Such questions as the creation of an oil company, delimiting its sphere of activity, and increasing its capital all required the approval of those ministerial bodies to which the oil industry was subordinated. Hence, to operate successfully, it was essential to have good connections in the tsarist government. Nobel understood this very well and, unlike other firms, he did not establish his headquarters in Baku (though he had an office there), but rather in St. Petersburg. By the mid-1880s Nobel had succeeded in forging strong ties to the apparatus of the tsarist government. He also had personal ties to a member of the tsar's own family, Grand Duke Mikhail, who for many years served as the emperor's governor-general (*namestnik*) in the Caucasus. Nobel came to Baku for the first time in 1876 (to investigate personally the potential of the oil business), and from Baku he went to Tiflis—the center of the state administration in the Caucasus—to meet the tsar's governor-general, who subsequently became a close acquaintance of Nobel. Without the personal connections in high places, which he regularly and expertly exploited, Nobel could never have achieved the position that he came to hold in tsarist Russia. In gaining access to these spheres, Nobel did not of course eschew the proven means of the bribe—in money, stocks, or "gifts." Rockefeller, as we have seen, also made ample use of graft, and the same methods were utilized in Russia.

At the same time, the tsarist government showed a desire to strengthen the position of the Nobel firm, which it saw as a reliable support for resolving questions that involved oil production and trade. Nobel himself, or his representatives, regularly participated in various government meetings. Indeed, virtually no decision on any question of significance was taken without first consulting with Nobel. The meetings were convoked sometimes at the initiative of the ministries, sometimes at the request of Nobel. In those cases when Nobel disagreed with a decision of the government, his vote could change that decision. Without Nobel's collaboration, the tsarist state would have had far greater difficulty controlling and regulating the petroleum business.

One of the government's main concerns was to impede penetration by Standard Oil (only later would the government approve the idea of an agreement with the American trust to divide up the world market). Nobel too regarded Rockefeller as a mortal enemy, but to strengthen his monopoly's position in Russia, he sometimes struck a deal with the American company—which provoked countless complaints from petty, middling, and even the larger oil industrialists. But Nobel came out unscathed, for the government closed its eyes to this flirtation with Rockefeller.

Nobel expertly combined his status and high connections with a merciless struggle against his rivals. He bought up oil tracts, cut prices, made loans on crushing terms, and when all this did not help, even resorted to criminal acts. For example, in the spring of 1885, after he had bought up a land plot that previously belonged to the city and was to become the site of a railway line, Nobel gave orders for a stone wall to be built in the path of the line and to scatter the stacks of rails and ties. According to the Council of the Baku Oil Producers Congress, Nobel's hired men had initiated "a real battle" and that "it was only possible to disperse them by [using] a company of cossacks specially sent for that purpose."[30]

In dealing with his competitors in Russia, Nobel acted as mercilessly as did Rockefeller in America. But the conditions under which he acted were substantially different from those in America. If in America capitalism had developed according to a classic paradigm, free from any kind of remnants from the historical past, in Russia things took a quite different course: the development of capitalism proceeded under conditions of a semi-feudal tsarist regime. Therefore, alongside advanced capitalist enterprises like Nobel's, there still existed backward business organizations like "business houses" (*torgovye doma*), which, because of certain traditions, remained quite strongly entrenched.

THE BUSINESS HOUSE OF A.I. MANTASHOFF

As a counterweight to the Nobel firm, the petty and middling firms in Baku formed a peculiar bloc under the aegis of local oil magnates—Mantashoff [Mantashev], Tagieff [Tagiev] and Gukasoff [Gukasov]. Insofar as they held a prominent role in the Baku oil business, Nobel had to deal with them. The leading position among these local enterprises belonged to the Business House of A. I. Mantashoff, whose head was typical of the Baku oil business at the time. The oil tracts that Mantashoff had acquired were very promising and quickly made him wealthy. The complete antithesis to a respectable capitalist like Nobel, Mantashoff had an office in Baku (with branches in other cities and countries), but ran his business at first in a very primitive fashion—in accordance with old merchant traditions. For example, when oil producers delivered the petroleum to his storage tanks, Mantashoff did not transfer payments through a bank but paid them off personally, sitting high upon his horse and pulling the money out of his pocket. He was miserly and calculating in money affairs, but in his personal life he conducted himself like an oriental monarch, surrounding himself with luxury and diversions; beautiful women from all over the globe were delivered to his residence. He enjoyed orgies of "Athenian nights," rolled out sumptuous banquets, with the table groaning under the weight of food and drink, as guests were entertained by troupes of dancers and acrobats. Mantashoff made journeys abroad; in Paris, Nice, Monte Carlo, and Cairo he was a constant visitor to casinos and houses of entertainment. This Croesus of Baku was poorly educated; he could not write properly in Russian and at first did not know foreign languages at all. Stories are told how, when he visited restaurants abroad, he would take a seat near the entrance to the kitchen and see what the waiters were bringing out, and then point with his fingers to make his order.[31]

For want of systematic statistical data on the condition of commercial and industrial groups engaged in the Russian oil business, it is not possible to determine in exact quantitative terms the relative position of Mantashoff and his allies. Nevertheless, by combining the data available in the scattered sources and the secondary literature one can draw some conclusions about the magnitude of their business operations. At the time when Nobel and the Rothschilds controlled approximately half of the oil production, two-thirds of the oil refining, and more than half of the petroleum trade inside Russia, and three-quarters of the kerosene exports, the share of the Mantashoff group constituted about one-third of the oil production, refining, and trade inside the country as well as a quarter of the kerosene exports.[32]

The semi-literate nouveau riche Mantashoff possessed practical skills and emerged as a leading entrepreneur. He began his business career in 1876 as a merchant in the first guild in Tiflis, and then began to engage in the oil trade in 1883. In 1886 he built a plant in Batumi to manufacture metal containers, and by the end of the 1890s it was producing 12 million containers a year. Mantashoff began to sell kerosene abroad in Turkey, Egypt, and Algiers, and opened his own commercial agencies in the Mediterranean ports of Constantinople, Smirna, Saloniki, Alexandria, Cairo, and Port Said, as well as Bender Abbasa in the Persian Gulf. From the early 1890s he also began to export kerosene to the Far East—to India and China—after opening agencies in Bombay and Calcutta, as well as using the services of the influential British firm of Jardine Mathieson & Co. In 1892 Mantashoff also opened an office in London.[33]

The marketing of oil products became his main business, although, right to the turn of the century, he also participated in a fishing company, dealt in real estate, and remained a share-holder of the Tiflis Commercial Bank. The scale of activities of the Mantashoff Business House, established only in 1892, grew so rapidly that even experienced officials in the government tax offices could not give any kind of exact figures on its size. "In general, the commercial activity of Manashev is so vast and so far-flung," wrote the head of the Tiflis office of the Treasury to his superiors in St. Petersburg, "that I would find it difficult to give a more or less accurate assessment of the dimensions of the turnover in his business activities." It was rumored that Mantashoff's profits amounted to 2 million rubles a year.[34]

By the time his firm was incorporated in 1899, Mantashoff owned oil refineries, pipelines, storage tanks and tankers for bulk shipping. The annual output of oil from his enterprise reached 800,000 tons. Mantashoff's port agencies in Batumi and Odessa were linked by rail lines, and some 100 of his own tank cars moved across the lines of the southwestern railway lines.[35]

During the reorganization of the business house, the former owners A. I. Mantashoff, M. A. Aramiants and others received 13.3 million rubles in stock and 6.7 million rubles in cash. When the joint-stock company was formed (with a capital of 22 million rubles), the leading banks of St. Petersburg and Moscow participated and held more than half of the stock at the founding session. A. I. Mantashoff and M. A. Aramiants held stock shares amounting to 10 million rubles, which they deposited in the Russian Bank for Foreign Trade as security for a loan of 9 million rubles.[36]

Relying upon the support of banks and holding the smaller and middling firms as allies, Mantashoff represented a separate force and was capable of preserving his independence—notwithstanding pressure from magnates like Nobel and the Rothschilds.

THE ROTHSCHILD BROTHERS IN BAKU: ATTEMPTS TO MAKE PEACE WITH ROCKEFELLER

The oil competition in Russia became more complicated after a representative from the Paris bank, "the Rothschild Brothers," appeared in Baku. This bank had enormous financial resources at its disposal and, as we have seen, at one time gave monetary support to the Nobel firm, thereby deriving an indirect profit from the Russian oil business. But in 1886 the Rothschilds purchased a Baku oil and commercial firm, which they renamed the Caspian-Black Sea Commercial and Industrial Co., with a capital of 1.5 million rubles.[37]

The Baku firm of the Rothschilds had stronger international ties than Nobel. At first, however, the Rothschilds did not want to engage in open conflict with Nobel. They therefore decided not to become actively involved in the domestic Russian market, where the Nobel firm was particularly strong. Originally, the Rothschilds concentrated their forces on the export trade in kerosene and, by the end of the 1880s, had achieved significant success and ranked as the leading exporter of Russian kerosene. To expand their influence in Baku, the French financiers made wide use of credit and loans, which—like Nobel—enabled them to draw a large number of middling and small firms into their orbit. Thus, the Rothschilds created a kind of bloc, upon which they could rely in their struggle against competitors both within Russia and in the foreign markets.[38]

The newspapers reported that the Rothschilds purchased the petroleum firm in Russia allegedly at the instigation of Rockefeller. It was even claimed that the Paris financiers were dummy front-men for the Americans. Neither report was true, but these attacks on the Rothschilds found fertile soil not only among their rivals in the oil business, but also in the gentry and landowner circles that were so ill-disposed toward foreign capital.[39]

To overcome this opposition, the Rothschilds sought support in the tsarist government and, like Nobel, solicited defenders inside the ministries. By the time the Paris bankers had entered the petroleum business, they already had some Petersburg connections at their disposal. Their main link was the prominent tsarist official, K. A. Skal'kovskii, who, as vice-director of the Mining Department in the Ministry of State Properties (from June 1885) and then its director (April 1891 to May 1896), was directly in charge of the oil business. By maintaining close relations with Skal'kovskii, the Rothschilds' representatives obtained needed information and made use of his support—which was virtually unlimited. And they paid Skal'kovskii handsomely, giving him money, presents, and information that enabled him to play successfully on the stock market.[40]

Skal'kovskii's position in high society and his behavior in the dealings with petroleum are quite remarkable. A highly placed tsarist official, he simultaneously participated in a newspaper in the capital (*Novoe vremia*), where he wrote about foreign policy and art (he was a fanatical devotee of ballet and the author of a book on ballet, in which he reported that he had attended performances of "Konek-Gorbunok" alone some 135 times). Skal'kovskii could assist the Rothschilds and their solicitors both because of his official position in the government and also because of his connections at court.

For his achievements in service Skal'kovskii was rewarded with the most distinguished imperial medals and titles. He willingly served those who came to him in the Mining Department through the backdoor, and in return he received due compensation for his efforts. In the words of the prominent journalist, L'vov-Kliachko, Skal'kovskii broke all records in graft: "On the one hand, his intimate connections with ballerinas (a significant number of whom were unofficial members of the imperial house)," wrote Kliachko, "and, on the other hand, his status as a collaborator in *Novoe vremia,* all that made him invulnerable." Kliachko goes on to add that "everyone who had business in the Mining Department knew that Skal'kovskii had to be paid off, otherwise nothing would be done." He mastered his role well and "all but advertised his bribery." A group of oil industrialists once petitioned Skal'kovskii to allot them a plot of state land containing oil reserves. In the event he satisfied their request, they would give him 20 thousand rubles and promised that "they will tell no one." Skal'kovskii replied: "Give me 40,000 and tell everybody." During his next drinking bout Skal'kovskii told this story himself, cynically declaring that "these fools" value "their silence" too highly.[41]

Extant documents from Skal'kovskii's personal correspondence reveal that he practically lived off the support of the oil industrialists. In a letter to Skal'kovskii in March 1886, Mantashoff noted that "now everyone is thinking about oil," and added that "I am very grateful to you for your attention, and for the fact that you do not forget me among all the profitable enterprises."[42] But the Rothschilds' representatives, M. Baer and J. Aron, virtually inundated him with various requests. Thus, with respect to the rules for shipping oil on the Trans-Caucasus Railroad, they asked Skal'kovskii to "arrange it such that no decision is taken which is contrary to our interests." "This will be all the easier, since our interests are identical with those of other industrialists, with the exception of N[obel]."[43] They also turned to Skal'kovskii to learn the conditions for impending sales of oil-bearing lands as well as changes being planned in the railway tariffs on petroleum shipping. They further solicited his assistance in obtaining permission to increase the capital of the Caspian and Black Sea Company (from 1.5 to 6 million rubles), which, under the rules of tsarist law, required the approval of the government. For a number of years the Rothschilds failed to obtain the necessary approval, but they finally solved the problem with the support of Skal'kovskii's good offices. "Our question about an increase in the capital is virtually resolved," wrote Baer, "and I am very grateful to you for this. Your order will be filled after 1 January."[44]

As one can judge from the correspondence with the Rothschilds' representatives, Skal'kovskii received monetary compensation from them each year. But if Baer's correspondence contains only allusions or hints of such payments, Mantashoff's letters call things by their real name: "I am enclosing, my dear Konstantin Apollonovich, two rugs. The next time we meet I shall tell you the reasons why you did not receive the rug I sent last year." And "I ask you to accept the enclosed caviar." And so forth.[45] Sometimes the bribes took the form of gifts to Skal'kovskii himself, sometimes as checks for gifts to women of the demimonde with whom Skal'kovskii had friendly relations. In Paris and on the Riviera, to which this tsarist official traveled each year to spend his holiday, he supported an entire harem of fashionable beauties with resonant names like Fufu, Margot, Manon, Rose-Leon, and so on. Mantashoff sent checks to buy gifts for them. "With the attached card, I ask you to receive 2000 francs from the Russian Bank [to cover] my debt to you for the present for Manon ... and another 1000 francs for [another] present for her." "I am attaching a check to the Russian Bank in Paris for 2000 franks, and request you to give this sum to Margot."—"1000 franks are for Fufu."—"I am attaching 600 franks to give to Rose-Leon."[46] Apart from the specifics in this correspondence, in essence they differ little from what one finds in Rockefeller's correspondence about the bribing of senators and other politicians in the United States.

Skal'kovskii's relations with the business world were not openly manifested. But some things did become known, and it was the threat of scandal that forced him to retire. After leaving state service in 1896, Skal'kovskii went straight into business enterprises, where he held the post of director in a number of Russian and foreign companies. In one of his newspaper articles he wrote: "I was and still am close to the industrial circles that invest their capital in Russia."[47]

Many representatives of the Russian aristocracy went abroad each year and spent their holidays in France. They formed a unique club of constant visitors to the Riviera, which included high tsarist officials and directors of a number of Russian banks and industrial firms, who were linked together not only by business but also by a mutual participation in the diversions and entertainment for which the French capital was famed. To make the stay as pleasant as possible, the Rothschilds spared no expense: they put their lodge at the opera at Skal'kovskii's disposal, organized receptions in his honor, and soon after his appointment as director of the Mining Department solicited for him the highest French honor—the medal of the Legion of Honor.

Not long before Skal'kovskii's resignation, the Rothschilds attempted (with his assistance) to recruit another high-ranking official in the government. They began with the director of the Department of Trade and Manufacturing in the Ministry of Finance, V. I. Kovalevskii, who controlled all matters connected with foreign institutions operating in Russia. In a letter to Skal'kovskii, M. Baer (director of the Caspian and Black Sea Company) expressed his confidence that "your colleague in trade and industry will be inclined to help you," and then proceeded immediately to raise the question whether Kovalevskii would not be willing to assume "the initiative in some kind of serious matter," that is, agree to carry out assignments like those given to Skal'kovskii. "Of course, it would all be a snap," M. Baer wrote a month later, "if Kovalevskii would want to take charge of this matter."[48] Evidently, however, nothing came of this venture.

The conditions of industrial and commercial entrepreneurship in Russia were alleviated in the early 1890s by important personnel changes in the tsarist government. At the end of 1892, S. Iu. Witte became minister of finance—a position which, in the previous decade and a half, had come to play a decisive role in questions of trade and industry. He was a defender of capitalist reforms and favored measures to attract foreign capital. As a result, despite legal limitations, foreign capital—with Witte's assistance—increasingly penetrated into Russian enterprises. It was no accident that the first news about Witte's impending appointment was greeted by the Rothschilds with great satisfaction.[49] Second only to this post, in terms of its impact on the interests of oil capitalists, was that of minister of state domains, to whose ministry the Mining Department was subordinated. And to this post was appointed A. S. Ermolov, whose views were very close to those of Witte. His predecessor, M. N. Ostrovskii, had opposed giving access to foreign capital. As M. Baer wrote to Skal'kovskii: "I assume that the situation has completely changed today, and that there will be no more difficulties."[50]

By forcing the development of Russia's capitalist relations, Witte hoped to expand its economic base and to strengthen the position of tsarism in the international arena. He encouraged oil entrepreneurs in their efforts to win markets for the sale of oil, supporting the export of Russian oil as it encountered opposition from America's Standard Oil.

By the early 1890s, the Russian petroleum companies had attained significant success on the world market. This was a result of the protracted and fierce competition with the Rockefeller company. At one time, in the late 1870s, American kerosene had been sold in Russia itself. According to Baron Rozen (the Russian general-consul in New York), no other country consumed such a quantity of American kerosene. In 1872-1873 the Americans exported 14.9 million gallons of kerosene to Russia, and in 1879-1880 7.9 million gallons. But thereafter, particularly after the establishment of the Nobel firm, the import of American kerosene into Russia fell sharply, constituting only 1.7 million gallons in 1882-1883 and then ceasing altogether.[51] The loss of the Russian market proved a severe blow for the Rockefeller firm. In 1881 Russian kerosene for the first time was exported (in a volume of 134,000 poods), and two years later it had risen ten-fold (to 1.5 million poods). In 1885 it amounted to 7.3 million poods, and after another three years had increased to 35.4 million poods.[52]

This explosive growth of Russian oil exports provoked concern among the leaders of the Rockefeller trust, and the Russian question became the subject of a special discussion. The heads of Standard Oil decided to take "prompt and energetic action" to halt this Russian invasion.[53] Of the planned measures, the most important was the creation of branches of Standard Oil, which, between the mid-1880s and early 1890s, were established almost everywhere in Europe—in England, Germany, Belgium, Holland, Italy, and other countries.

An Industrial Commission of the United States, appointed in 1898, studied the prospects for expanding American exports and recommended that a network of banking agencies be established "in the leading centers of world trade."[54] Standard Oil was ahead of that recommendation, for by the early 1890s it had already established its own branches in the above countries by buying up local firms engaged in the kerosene business. The branches remained nominally independent, but in reality were firmly in the hands of Standard Oil. In England, for example, the entire stock of the branch (worth 2.5 million dollars) belonged to Rockefeller. In Belgium he owned 51 percent (out of a 2 million dollar capitalization) and in Italy 60 percent (out of a capital of 500,000 dollars).[55] These companies were organized like the marketing divisions of Standard Oil in the United States, with their own means of transportation, storage facilities and everything necessary for the sale of American oil products. Two decades later, in 1907, Standard oil already controlled 55 foreign branches, whose aggregate capital amounted to 37 million dollars.[56]

Almost simultaneously, Nobel and the Rothschilds began establishing branches in European countries. The Rothschilds opened theirs in England, Belgium, and Holland. The Nobel firm opened relatively small offices in these countries, but devoted its main attention to organizing branches in Austria and Germany, which were to become its prime base of activity in Europe. Like Standard Oil, the Russian companies owned the controlling interest of stock in their branches, but the total sum of their investment was considerably smaller than that of the Americans, less than four to five million dollars (in comparable figures).

Over the course of several decades the world economy and politics proved to be in the domain of multi-national corporations. However, it was precisely the oil companies that laid the foundations for this phenomenon and provided the origins of contemporary forms of monopolistic capital. The structure of the first oil trusts and their international organization, in many respects, reminds one of contemporary multinational corporations. Hence the oil companies of that period, with their numerous foreign branches, can be seen as the prototype of modern multinational aggregates.

Simultaneous with the establishment of branches in West European countries, Rockefeller's firm decided to embark on negotiations with the oil companies in Russia, beginning with Nobel. In the event of an agreement, it expected to subordinate other oil producers to itself or, at least, to exacerbate the tensions already prevailing among them. Aware that the Nobel Company needed financial support, Rockefeller proposed to provide financial aid in order to obtain "control of the business in Russia" and "to absorb the Nobels." Reports to this effect appeared in the American and then the Russian press.[57] In the opinion of the Russian consul in New York, Baron Rozen, such newspaper reports were not without foundation.[58] This was subsequently confirmed by archival materials. Documents in the Rockefeller archive demonstrate that the heads of Standard Oil did consider the question of purchasing the Nobel firm.

During a business trip to Europe in 1883, one of the directors of the Rockefeller foundation—W. Warden—held discussions in Paris with "an Admiral in the Russian Navy with a great long name," who spoke on behalf of the Nobel Brothers about the possibility of "some arrangement with the Standard Oil Company."[59] In a more detailed form, the same subject was raised in the correspondence between Warden and the representative of a British commercial firm, G. Funk, a contractor of Standard Oil. They discussed the proposal for an agreement with Nobel, which presumed that the American company would acquire a large share of the Nobel stock with the goal of joint collaboration in the construction of a Trans-Caucasus pipeline.[60]

Funk regarded the scheme to purchase the Nobel firm "impracticable" in view of the inevitable opposition of the tsarist government. He noted that "it is necessary to keep L. Nobel largely interested and active in the concern on account of his capacities, his knowledge, and his experience not only as regards the business, but also in dealing with the bureaucracy of an autocratic Government, and his high connections."[61] Nothing came of this proposed agreement with Nobel. In his report to Rockefeller, Warden explained that he never really had any great hopes of succeeding. But he simply wanted to clarify the question with Funk in order to have "all the information in the future not only in relation to the oil fields in Russia, but also in relation to the financial standing and ability of these men who are engaged in this business.[62]

The Parisian Rothschilds also became the object of intent interest as soon as they founded their business in Baku. Suspicions even arose that Rockefeller and the Rothschilds were collaborating. Although, as already noted, such assertions were groundless, Rockefeller's representatives did try to make approaches to the Rothschilds. Standard Oil first discussed the possibility of an agreement with the Rothschilds as early as 1886, that is, just one year after the Caspian and Black Sea Company had been founded.[63] In the following years representatives of the American oil trust journeyed repeatedly to Paris to hold discussions on this issue. Emissaries of Rockefeller often visited Baron Alphonse, the head of the Rothschilds' enterprise. "Baron Rothschild speaks English fluently," noted Standard Oil's manager, John Archbold.[64] Admittedly, Alphonse's son-in-law, Jules Aron, the firm's chief agent for the oil business, did not know English, but Rockefeller's representatives nonetheless found it possible to hold productive talks with him.

As a result of prolonged negotiations between Standard Oil and the Rothschilds, a draft proposal was finally worked out. When Archbold reported about this to Rockefeller, the latter gave his go-ahead: "I think you are doing right in reference to the Russian policy."[65] The two sides had reached an agreement to divide up the sphere of influence on the world market, which would take effect under two conditions: (1) if it were signed by other Russian oil producers, in the first instance, the Nobel firm; (2) if it were sanctioned by the tsarist government, which would guarantee its support. The Rothschilds kept the Ministry of Finance informed on the course of the negotiations. The ministry did not decline to support this kind of agreement, but it did make some reservations. The minister of finance, Witte, held that, for the success of Russian oil exports and an advantageous agreement with America, it would be necessary to make a preliminary union of Russian oil exporters into a syndicate, operating under the supervision of the government.[66]

This step was regarded as a practical measure in realizing its economic program. But such an objective proved difficult to achieve, for the Russian oil producers were divided by fundamental differences, which had substantially intensified by the early 1890s. This was true, above all, for the relations between the Nobel and Rothschild firms.

Having lost the possibility of obtaining finances in Paris, Nobel turned to Berlin—a highly revealing act in itself, considering the tensions in Franco-German relations. As already noted, the company had originally enjoyed the financial assistance of the Rothschilds. It also received money from the Crédit Lyonnais. But it thereafter refused the services of the French bankers and turned to the German Disconto Gesellschaft, and the Berliner

Handelsgesellschaft for credits. The German banks began to supply Nobel with funds and continued to finance the firm for the next three decades, up to the very outbreak of World War I. In 1884 they gave Nobel two loans for 4 million rubles, and in 1889 provided another loan for 2 million rubles,[67] creating a syndicate to place the Nobel securities on the Dutch, German and Belgian stock markets.[68] The St. Petersburg International Commercial Bank, which was closely associated with the Disconto Gesellschaft, played an important role as intermediary between Nobel and the German financiers.[69] Then, and in later years, the St. Petersburg bank actively assisted in attracting foreign capital to Russia.[70]

The German banks earned fat profits on their dealings in Nobel securities, obtaining good commissions from the stock market sales. In 1884-1894 their portfolio contained a large number of stocks, constituting 9 percent of their entire fixed capital.[71] The real value of these shares was actually much greater, for the stock-market rating was significantly higher than their nominal value.

In turning regularly to the German banks for financial assistance, the Nobel company did not, however, become dependent upon them. The Russian firm paid off its debts punctiliously and maneuvered adroitly in its relations with the different banking houses. It was assisted here by its connections with the St. Petersburg International Commercial Bank and the Volga-Kama Bank, the Swedish Ensckilda, and other banks.

The strengthening of Nobel's financial position forced its chief Russian rival, the Rothschilds, to take measures to consolidate their forces. In the face of the deteriorating relations with Nobel, the Rothschilds sought to create a counterweight by forming an alliance with British firms. The alliance was facilitated by their mutual interests in the oil trade in the Asia, where the English had also become active since the late 1880s.

The petroleum business was intimately related to the general political interests of England and, in particular, to the Anglo-Russian relations in this region of the world. The kerosene trade could prove a major factor in determining a country's political influence in the region. England aspired to be the master of the Eastern market and feared the competition of other powers and above all Russia, which it regarded as its traditional rival. Meanwhile, the marketing of Russian oil in India, Japan, and

China had already become entirely competitive, especially after the opening of the Suez Canal, which made Russia's route to the East shorter, and thereby enabled the Russian companies to challenge Standard Oil's hegemony.

The prospect of such a development aroused unconcealed alarm in England and impelled the English to make an attempt to put the Russian petroleum business in Asia in their own hands. In the mid-1880s the firm of Lane and McAndrew established relations with Nobel. Soon, however, Nobel spurned its services and sought another contractor—the firm of Bessler, Wächter, and Co., which, beginning in 1888, became its branch office in London. Then Lane and McAndrew entered into contact with the Rothschilds, and in 1888 founded in London a joint firm, Kerosene Co. Frederick Lane, the head of the British firm, played an important role in making this deal, and he later proved to be one of the most adroit diplomats in the oil business.

In 1886 Lane and MacAndrew exported the first shipment of Russian kerosene to Calcutta, and thereafter made the passage of ships to India, China, Japan, and other countries to the East part of their regular operations. In the late 1880s the English company began to import bulk shipments of oil to the East in tankers, which enabled a significant increase in petroleum deliveries.

An alliance with British capital became the cornerstone of the Rothschilds' policy. The agreement with Lane and MacAndrew was supplemented in 1891 by a pact with the famous British commercial firm of Samuel, Samuel, and Co., which became active in the oil markets. As a result of its agreement with the Rothschilds, the Samuel, Samuel, and Co. took control of approximately one quarter of Russian oil exports and obtained monopoly rights to sell Russian kerosene in the entire region east of the Suez Canal.[72]

With the assistance of the French representatives to the administration of the Suez Canal, the Rothschilds obtained official permission for their oil tankers to pass through the canal. News of this at first provoked sharp protests in England, which had acquired the controlling share of stocks in the Suez Canal Co., and virtually determined its activity. English steamship companies sent the British government petitions of protest against the permission given to the Rothschilds and proposed that this question be considered by the government commission on naval affairs. They fabricated a conclusion of experts alleging that the transit of tankers through the canal would pose a threat to other shipping. The press and interested parties demanded that the government have the permission of the Rothschilds abrogated.[73] Nevertheless, the permission remained in force, for a powerful British corporation—the Samuels' company—had a vested interest in the export of Russian oil to Asia and supported the Rothschilds. In January 1892 the policy became universal, when the Suez Canal published a decree authorizing oil ships to pass through the Suez, thereby providing Russian oil access to the Asian markets.

The Samuels firm headed a large British syndicate that combined a number of business houses in India, Japan, China, and other countries. It had been engaged in trade in the East for several decades, and the fact that it now controlled the sale of Russian kerosene was fully acceptable to imperialist circles in Great Britain.

The Nobels' firm—Bessler, Wächter and Co.—had tried to prevent the agreement between the Rothschilds and Samuels. It appealed to the tsarist government to oppose the movement of oil tankers through the Suez Canal, explaining that this "threatens to deal a serious blow" to the interests of Russian exporters.[74] Although the Nobel company itself had initiated the bulk shipment of oil by tankers on the Caspian Sea and the Volga, and later in the Baltic, it now protested when its rivals began to use the very same techniques. It argued that the tanker trade through the Suez could damage the canal and thereby inflict harm to Russian interests. However much the Nobel firm might dislike the agreement between the Rothschilds and Samuels, it was unable to undermine it. At the next meeting of the Baku oil producers Nobel tried to obtain a resolution against the tanker shipment of Russian oil to the East, but nothing came of its efforts. The arguments of Nobel's represented were judged to be "groundless." Skal'kovskii summed up the results of the discussion at the meeting: "All the firms not only do not share the fears expressed by Bessler, Wächter and Co., for the fate of the Batumi and Baku oil industry, but, on the contrary, see the permission for bulk oil tankers to pass through the Suez Canal as ensuring further development of this industry."[75] Thus Nobel's effort to deal a blow to the Rothschilds' agreement with the British firm of Samuels ended in utter failure.

PROPOSALS TO DIVIDE THE WORLD MARKET

As a result of the growing economic significance of oil and the appearance of powerful corporations, by the end of the nineteenth century the oil business had attained a large scale and exerted a major influence on the economic and political development of countries and international relations. A kind of empire was created, with constant competition among various centers generating recurrent crises and conflicts over oil. But periodically the oil companies made an effort to reach a peaceful agreement to divide up spheres of influence and put an end to the conflict.

At the outset of the 1890s there was an effort to create an export syndicate of Russian kerosene manufacturers. Its fundamental premise was that the world market would be divided between the Russian oil producers and Standard Oil of America.

The Russian Ministry of Finance, however, looked askance at such schemes. Thus, although negotiations to divide spheres of influence on the world market began in the late 1880s and actually reached a preliminary understanding in the early 1890s, the Ministry of Finance stepped in to quash the scheme. In the Ministry's view, before Russian oil producers make a pact with the Americans, they should first reach an agreement among themselves in order to enter the world market as a united front. This goal proved more logical than attainable. However, in the face of powerful pressure from Standard Oil, which was in turn supported by the United States Government, the Russian Ministry of Finance decided to take measures in response.

The well-known American historian, Allan Nevins, has noted that the Rockefeller oil trust, while subject to government limitations inside the United States, enjoyed its unqualified backing on the world market.[76] The official diplomatic service worked in close contact with representatives of Standard Oil. Rockefeller himself later admitted: "One of our greatest helpers has been the State Department in Washington. Our ambassadors and ministers and consults have aided to push our way into new markets to the utmost corners of the world. They appreciate the value of going hand in hand with our agents."[77] This quotation is taken from Rockefeller's memoirs. To be sure, the last phrase was removed from a later edition, but is found in the original, located in Rockefeller's personal archive.[78] For all practical purposes, Standard Oil enjoyed the unlimited support of the American government.

The Russian Ministry of Finance studied the American experience carefully in order to make use of this in Russia and to strengthen the position of the Baku firms on the world market. In dispatching I. I. Ianzhul (a professor at Moscow University and prominent Russian economist) and S. I. Gulishambarov (a leading expert on oil questions in the Ministry of Finance) to the World's Fair in Chicago in early 1893, Witte gave them the goal of studying the organization of American trusts and syndicates, above all that of Standard Oil. Their reports were published and provided the point of departure for a decision about the necessity of creating a Russian syndicate for the export of oil. Thus, what had unfolded in America as a natural process became a different process in tsarist Russia, where the government decided to impose this "from above," by means of state intervention.[79]

Continuing to hold the reins of power in his own hands, in August 1893 Witte ordered that an invitation be sent to representatives of the pertinent ministries to participate in a special conference, convoked jointly by the ministry and petroleum industrialists. Its goal was to discuss the conditions for creating a syndicate of kerosene manufacturers so as to conduct "more successful resistance to Standard Oil on international markets."[80] The proposal to organize the syndicate was prepared, with the approval of the Ministry of Finance, by the Nobel and Rothschild firms. They wanted to create something similar to the association of kerosene manufacturers, which Rockefeller had created earlier in the United States. The Nobel and Rothschild proposal foresaw the establishment of a Union, which would regulate the level of prices and volume of deliveries; the quota for each firm participating in the agreement would be determined by its productivity and volume of commercial activity in the previous year. Control to see that the terms of the agreement were observed was to be assigned to a committee of the Union and its commercial agents. But since only those "who have an existing organization for the sale of kerosene on foreign markets" could serve as commercial agents, this meant that the affairs of the future association would rest entirely in the hands of Nobel and the Rothschilds, for these were the only firms that had their own branches abroad. This was confirmed in a statement on October 2, 1893 by the director of the Department of Trade and Manufacturing, V. I. Kovalevskii, whom Witte had given the task of running the work of the conference. In characterizing the "essence" of the proposal under discussion, Kovalevskii announced that "the sale of kerosene abroad is delegated exclusively to those firms, which have a properly organized business there." The government made clear its interest in having the agreement signed by all the kerosene manufacturers, promising to grant a preferential railway tariff and other assistance to members of the Union once it had reached an agreement with Standard Oil to divide up the world market.[81]

The conference lasted three weeks and ended in the signing of an agreement, which was then approved by the Ministry of Finance. Two-thirds of the Baku kerosene manufacturers signed the agreement. Although the Mantashoff group refused to sign, both Nobel and the Rothschilds believed that Mantashoff and his allies would be forced to join the agreement, for otherwise they would forfeit the special privileges being conferred by the government. The newspaper, *Kaspii,* wrote that the preferential treatment proposed by the government would mean that "it will be almost impossible to conduct foreign trade without joining the Union." Nevertheless, Mantashoff did not wish to surrender voluntarily.

In this difficult hour Mantashoff sent a personal letter to his protector, the Director of the Mining Department, K. A. Skal'kovskii, with a request for assistance. Mantashoff complained about what Nobel and the Rothschilds had done: "Having composed an agreement to their own taste and appointing themselves as its sole agents, they [seek to] attract other plant owners and to force them to sign the agreement by conveying the idea that the government has promised to reduce the railway tariff from Baku to Batumi only for this 'Union,' while plant owners who do not join the 'Union' will remain under the existing tariff." Mantashoff declared that they are either forcing him to surrender "to the caprice" of the syndicate's bosses, or to liquidate his enterprise. "These accursed kerosene affairs with Nobel and the Rothschilds have poisoned my life. Can they really force me to enter into a marital union when I do not wish this?"[82]

It is not known how Skal'kovskii replied to Mantashoff or whether he in fact did anything on behalf of the latter. In any event, it soon became known that the government privileges would take effect only after the Union included at least 85 percent of all the plant owners who were exporting kerosene. In other words, without the support of Mantashoff's group, it was impossible to count on receiving the government concessions. On the other hand, the government announced that it would approve the creation of a second autonomy group of the Union, if it signs an agreement with the first group and if they jointly enter into negotiations with Rockefeller.

This condition was fulfilled. In November 1893 the Mantashoff group signed an autonomous agreement, with Mantashoff designated as the commercial agent for the second group. Then, in February 1894, both groups reached an agreement on the joint sale of kerosene abroad and the division of the markets. The Mantashoff group received 66 percent of the kerosene trade in the Near and Middle East, in Africa and in the Balkans as well as 25 percent of the market in the Far East. Nobel and the Rothschilds secured the European market for themselves. A few days after the two groups had signed the agreement, the government introduced a preferential tariff on the Trans-Caucasus Railroad, with the reservation that in the future it will "take one or another measure depending upon the agreement with the Americans."[83]

Negotiations, conducted in the name of the Baku Union of Kerosene Manufacturers, were subsequently resumed in Paris with Rockefeller's representatives. A new preliminary agreement was reached on a division of spheres of influence and allocation of corresponding quotas for the kerosene trade. The agreement, however, required Witte's approval before it could go into effect, and the finance minister refused to give his endorsement. In the opinion of the American general-consul in St. Petersburg (John Carel), Witte rejected the agreement on the grounds that, given its terms, "the Russian kerosene producers may possibly receive, for the time being, some profit from the bargain, but that they would give themselves over into the hands of the Americans, and that the independence of the Russian naphta industry, which has a great future, would suffer therefore."[84]

Although Witte vetoed an agreement with Standard Oil, he did not reject the idea of an agreement with the Americans, and the negotiations continued. The Ministry of Finance helped to encourage the Union to make further efforts when it granted a significant reduction in the tariffs for the Trans-Caucasus Railway (effective July 1, 1894). The American consul in Batumi, John Chambers, who simultaneously was an agent of Standard Oil, commented on this measure with unconcealed irritation. In his words, the Russian representatives in Paris were deliberately procrastinating in the negotiations and bombarding Petersburg with telegrams, which spoke of the favorable prospects for an agreement with the Americans in the event of "further reductions in the freight rate would surely settle the affair promptly."[85]

A certain period of time passed, and reports circulated once more that an agreement had been signed with Rockefeller. And again the Ministry of Finance judged its terms to be unsatisfactory. Evidently, Witte had lost faith in the possibility of reaching an agreement, and he now declared that "he does not have any sympathy for this agreement and does not see the need for cooperative action of our oil industrialists with Americans on the basis of the terms that have been worked out."[86]

The agreement of the "Union of Baku Kerosene Manufacturers" was renewed in May 1895. The question of an agreement with Rockefeller was removed, although, because of the mounting tension on the world market, the problem of the relation of Russian oil industrialists to Standard Oil remained. But hopes of reaching agreement on honorable terms were not gone.

The syndicate of Baku kerosene manufacturers proved vulnerable to insuperable internal contradictions and, as a result, folded after less than three years of existence. As was explained years later by K. V. Gagelin (then the directory of the Baku branch of the Nobel Brothers), after two years and eight months of "onerous and tireless work," the Union disintegrated and the committee that controlled his activity ceased to exist. As a result, he wrote, "it became easier for exporters to work." At a dinner organized in Baku in honor of this occasion, G. Tagieff (the head of one of the largest local firms)—who had in the beginning been regarded as "a powerful supporter of the Union"—gave a speech in which he explained "why he changed his attitude."[87] Many representatives of local firms disliked the fact that they were subordinated to the syndicate's leaders—Nobel and the Rothschilds, although the latter two parties as well as the syndicate had turned into a burden. The agreement that originally regulated the export trade was no longer to their liking.

Although the tsarist government in the future would seek means to unite the oil industrialists into a common front against the competition of the American oil trust, nothing came of these efforts. A final attempt was made shortly after the turn of the century, but it too ended in failure.

Henceforth tsarist authorities decided that the only feasible course was to rely on the large firms, and indeed its main support continued to be the Nobel firm. Witte's orientation toward attracting foreign capital also provided favorable conditions for the Rothschilds and other financial groups with an interest in the Russian oil business. The ban on capital with ties to the Rockefeller oil trust remained categorical; a similar ban did not exist for other commercial associations. So far as their admission to the Russian oil industry is concerned, the rules on their access were determined by economic, as well as domestic and foreign policy factors.

NOTES AND REFERENCES

1. *Fortune,* October 1951, p. 99.
2. Quoted in D.F. Hawke, *John D. The Founding Father of the Rockefellers* (New York, 1980), p. 152.
3. Ibid., pp. 152-153.
4. Ibid., p. 144.
5. J. Abels, *The Rockefeller Billions* (New York, 1965), p. 59.
6. Ibid., p. 50.
7. J.D. Rockefeller, *Random Reminiscences of Men and Events* (New York, 1909), p. 12.
8. A.L. Moore, *John D. Archbold and the Early Development of Standard Oil* (New York, n.d.), pp. 115-121.
9. H. O'Connor, *World Crisis in Oil* (London, 1963), p. 61.
10. J. Moody, *The Truth about Trusts* (New York, 1904), pp. 110-111; F.A. Shannon, *America's Economic Growth* (New York, 1947), p. 424.
11. D.F. Hawke, p. 165.
12. Abels, p. 237.
13. Ibid., p. 236.
14. J.D. Rockefeller to V. Cary, December 22, 1916 (Rockefeller Archive Center [hereafter RAC]).
15. Abels, p. 237.
16. H.F. Williamson and A.R. Daum, *The American Petroleum Industry* (Evanston, 1959), *1*, p. 492.
17. I.A. D'iakonova, *Nobelevskaia korporatsiia v Rossii* (Moscow, 1980); John P. McKay, "Enterpreneurship and the Emergence of the Russian Petroleum Industry, 1813-1883," *Research in Economic History, 8* (1982), pp. 58-65.

18. *Monopolisticheskii kapital v neftianoi promyshlennosti Rossii. 1883-1894. Dokumenty i materialy* [hereafter *MKNPR*] (Moscow-Leningrad, 1961), p. 662.

19. See: V.A. Nardova, *Nachalo monopolizatsii neftianoi promyshlennosti Rossii. 1880-1890-e gody* (Leningrad, 1974); McKay, pp. 65-76.

20. Ibid., pp. 9-11. See also D'iakonova, pp. 59-60; R.W. Tolf, *The Russian Rockefellers* (Stanford, 1976), pp. 50-67; McKay, pp. 67-71.

21. *MKNPR,* p. 751.

22. D'iakonova, 140-43.

23. See appendix, Table 1.

24. See appendix, Table 2.

25. *MKNPR,* p. 751.

26. A.A. .Fursenko, "Mozhno li schitat' kompaniiu Nobelia russkim kontsernom?" In *Issledovaniia po sotsial' no-politicheskoi istorii Rossii* (Leningrad, 1971), 352-361; D'iakonova, pp. 71-75.

27. Tolf, p. 82.

28. Zhurnal zasedaniia Soveta Gosudarstvennogo Banka 29 iiulia (11 avgusta) 1905 (*MKNPR,* p. 341).

29. Nardova, p. 108.

30. *MKNPR,* p. 55.

31. R. Hewins, *Mr. Five Per Cent. The Story of Calouste Goulbenkian* (New York, 1958), p. 49.

32. *Obzor bakinskoi neftianoi promyshlennosti za 1901 g.* (Baku, 1902), chast' l. 431; chast' 2, l. 10, 213, 226, 230; *Obzor bakinskoi neftianoi promyshlennosti za 1914 god* (Baku, 1915), chast' 1, ll. 8-9; *MKNPR,* pp. 231-233, 752-754; Nardova, p. 140; L. Eventov, *Innostrannyi kapital v neftianoi promyshlennosti Rossii* (Moscow-Leningrad, 1925), pp. 39-41; S. Pershke and L. Pershke, *Russkaia nevtianania promyshlennost', ee razvitie i sovremennoe polozhenie v statisticheskikh dannykh* (Tiflis, 1913), 191-193.

33. Spravka Ministerstva torgovli i promyshlennosti ("Svedeniia ob eksportnoi torgovle kerosinom A. I. Mantasheva za 16 let, 1886-1901 g." ot 17/30 ianvaria 1901 g.). In *TsGIA SSSR,* f. 23, op. 24, d. 1066, l. 139.

34. Upravliaiuschichii Tiflisskoi kazennoi palatori—V.I. Kovalevskomu (31.III/13.IV.1900) in ibid., l. 136.

35. Ibid.; "Poiasnitel'naia zapiska uchreditelei k ustavu obshchestva 'A.I. Mantaschev i Ko.'" (12.24.I.1899) In *MKNPR,* pp. 231-233.

36. "Protokol zasedaniia uchreditel'nogo sobraniia A. I. Mantashev i Ko." (29.VI/11.VII.1899); spisok aktsionerov kompanii (29.VI/11.VII.1899); Otdelenie Volzhsko-Kamshogo banka—A. I. Mantashevu i Ko. (20.IX/2.X1899). In TsGIA SSR, f. 23, op. 24, d. 620, 1. 11, 70, 74, 117.

37. *MKNPR,* p. 665.

38. *MKNPR,* pp. 118-121.

39. *MKNPR,* p. 671.

40. A. A. Fursenko, "Materialy o korruptsii tsarskoio biurokratii," *Issledovaniia po otechestvennomu istochnikovedeniiu* (Moscow-Leningrad, 1964), pp. 149-155.

41. L. L'vov (L. M. Kliachko), *Za kulisami starogo rezhima. Vospominaniia zhurnalista* (L. 1926), pp. 87-88.

42. A. I. Mantashoff to K. A. Skal'kovskii (11/23.III.1886). In LOII, f. 202, d. 25, kart. 5, l. 58.

43. M. Baer to K. A. Skal'kovskii (13.25.III.1890). In LOII, d. 21, kart. 3, l. 167.

44. M. Baer to K. A. Skal'kovskii (28.XII.1895/9.I.1986). In LOII, ll. 252-253.

45. Mantashoff to K. A. Skal'kovskii (13/25.X.1892; 28.I/9.II. 1894;; 24.I/ 5.11.1895). In LOII, d. 21, kart. 5, 11. 60, 64 ob.

46. K. A. Slal'kovskii's personal correspondence has been broken up and is preserved in two archives: LOII and IRLI.

47. *Novoe vremia,* November 1, 1898.

48. M. Baer to K. A. Skal'kovskii (28.XII.1895/9.I.1896). In LOII, f. 202, d. 21, kart. 3, 11. 252-253.

49. M. Baer to K. A. Skal'kovskii (8/20.VIII.1892). In LOII, k. 209.

50. M. Bear to K. A. Skal'kovskii (28.XII.1895/9.I.1986). In LOII, ll. 252-253.

51. Report from Rozen (dated 25.III/6.IV.1886). In *TsGIA SSSR,* f. 574, op. 4, d. 315, ll. 208-209.

52. S. and L. Pershke, *Russkaia neftianaia promyshlennost', ee razvitie i sovremennoe polozhenie v statisticheskikh dannykh* (Tiflis, 1913), pp. 29, 39. In 1888 Russian kerosene exports composed 22 percent of world. exports. The remaining 78 percent came from the United States, of which nine-tenths belonged to Standard Oil (see R. and M. Hidy, *Pioneering in Big Business* [New York, 1955], p. 132).

53. A. Nevins, *A Study in Power. John D. Rockefeller, Industrialist and Philanthropist* (New York, 1953), *2,* p. 115.

54. *Final Report of the Industrial Commission,* vol. 19 (Washington, D.C., 1902), pp. 579-580.

55. Hidy, pp. 144-154, 533-546.

56. M. Wilkins, *The Emergence of Multinational Empires* (Camgridge, 1970), pp. 83-84.

57. *New York Times,* April 5, 1886.

58. Report from Rozen (dated 25.III/6.IV.1886). In *TsGIA SSSR,* f. 574, op. 4, d. 315, l. 179.

59. W. Warden to Rockefeller, November 9, 1884, in RAC.

60. Nevins, *2,* pp. 115-116.

61. H. Funk to W. Warden (October 26, 1883); in RAC.

62. W. Warden to Rockefeller (November 4, 1888), in RAC.

63. J. Archold to Rockefeller (April 3, 1886), in RAC.

64. Archbold to Rockefeller (July 13, 1883), in RAC.

65. Rockefeller to Archbold (June 14, 1889), in RAC.

66. See A. A. Fursenko, "Pervyi neftianoi eksportnyi sindikat v Rossii," in: *Monopolii i inostrannyi kapital v Rossii* (Moscow-Leningrad, 1962), pp. 4-58.

67. L. Nobel to the Petersburg office of the State Bank (June 8/21, 1884); E. Nobel to V. A. Liaskii of the St. Petersburg International Commercial Bank (June 19, and July 1, 1889). In *TsGIA SSSR,* f. 588, op. 1, d. 121, 11. 95-96; f. 626, op. 1, d. 121, 1. 41a.

68. J. Mai, *Das deutsche Kapital in Rußland, 1850-1895* (Berlin, 1970), p. 177.
69. Adolf von Hansemann (Disconto Gesellschaft) to Liaskii (June 9/21, 1884, April 20/May 2, April 26/May 8, July 2/14, 1889). In *TsGIA SSSR*, f. 626, op. 1, d. 107, l. 1; d. 121, 11. 391, 45a, 87.
70. S.K. Lebedev, "Peterburgskii mezhdunarodnyi kommercheskii bank v konsortsiumakh po vypusku chastnykh zheleznykh zaimov 1880-kh—nachala 1890-kh gg." and S.K. Lebedev and B.V Anan'ich, "Uchastie bankov v vypuske obligatsii rossiiskikh zheleznodorozhnykh obshchestv (1860-1914 gg.)," *Monopolii i ekonomicheskaia politika tsarizma v kontse XIX - nachale XX v.* (Leningard, 1987), pp. 5-64.
71. Mai, p. 177.
72. A copy of this secret agreement (from November 8, 1891) was subsequently obtained by the Nobel firm; see *TsGIA SSSR*, f. 1458, op. 1, d. 1355, 11. 1-5.
73. F.C. Gerretson, *History of the Royal Dutch* (Leiden, 1958), 1, p. 216.
74. A. S. Ermolov to Veshniakov (October 10/22, 1892), in *TsGIA SSSR*, f. 37, op. 66, d. 2176, ll. 1-2.
75. K. A. Skal'kovskii to Veshniakov (November 20/December 2, 1892) in ll. 1-2.
76. A Nevins, *John D. Rockefeller. The Heroic Age of American Enterprise* (New York, 1940), *2*, pp. 34-35.
77. Rockefeller, p. 63.
78. Folger to Jenkins (July 6, 1908); J.D. Rockefeller, Jr. to Muprhy (December 1908) in RAC.
79. A. A. Fursenko, *Pervui neftianoi eksportnyi sindikat v Rossii*, p. 5.
80. Witte to D.S. Sipiagin (August 31/September 12, 1893). In *Materialy po istorii SSSR*, vol. 4 (Moscow, 1959), p. 72.
81. Fursenko, *Pervyi sindikat*, pp. 12-13.
82. Mantashoff to K. A. Skal'kovskii (October 29/November 10, 1893). In LOII, f. 202, d. 21, kart. 5, ll. 62-63.
83. *Kaspii*, March 1, 1894.
84. Report from Carel (April 30, 1895). In *U.S. Consular Reports, 48*, p. 322.
85. Report from Chambers (March 1, 1895). In *U.S. Commercial Relations, 1894-1895*, 2, 320.
86. *Kaspii*, April 23, 1895.
87. K. V. Gagelin [Hagelin], *Moi trudovoi put'* (New York, 1945), pp. 249-250.

Chapter II

The Entry of Foreign Capital into Russia

The oil rivalry developed in the context of international relations that were to become ever more tension-filled. By the eve of World War I, it had become inextricably intertwined with events that would trigger military conflict. But even at earlier stages, the battle for petroleum already bore an important political undercurrent. A graphic example of this tendency was the struggle of international capital for control of Russian oil at the turn of the century.

The war between America's Standard Oil and Russia's producers and exporters would long continue to form the pivotal point in international petroleum competition. That conflict was aggravated as new capitalist groups entered the marketplace.

As published data (in millions of barrels) show,[1] between 1898 and 1901 Russia steadily widened its lead over the United States in the total volume of petroleum production:

	1898	*1899*	*1900*	*1901*
RUSSIA	66.60	66.00	75.78	85.17
USA	55.36	57.07	63.62	69.39

By the turn of the century the Russian oil industry proved vulnerable to pressure from numerous foreign companies that were seeking to obtain various concessions. The intrusion of foreign capital into Russia was one of the most telling manifestations of international oil competition in those years. The conditions for obtaining commercial access to Baku, then to the new oil-producing regions (Groznyi and Emba) proved rather favorable, notwithstanding the restrictions imposed by tsarist legislation on industrial institutions, especially when foreign companies were involved. As already pointed out, S. Iu. Witte's tenure as minister of finance coincided with a new policy of deliberately seeking to attract foreign capital, whose assistance Witte deemed vital to the realization of his program of economic transformation.

For many countries of the world, development of the economy and (especially) industry proved to be closely tied to the participation of foreign capital. It is well known, for example, that the industry and railways of the most developed country, the United States, were constructed, in large measure, on the basis of European (primarily British) capital. Moreover, the volume of foreign capital (3 billion dollars in 1890) and its influence on the country's development was so great that, notwithstanding its enormous economic growth, for a long time the United States was still economically dependent on Europe. To be sure, certain important branches of industry, including petroleum, developed almost entirely on the basis of domestic capital, which served to moderate foreign influence. For this and other reasons, the United States' economic dependence on European capital did not prevent that country from pursuing an independent policy, even if it did, from time to time, impede its activities on the world arena.

But the situation was quite different when foreign capital was exported to backward countries or colonies. Although the foreign investment contributed to the growth of capitalism, its development was slow and one-sided and also led to an intensification of economic and political dependence. That is precisely how things developed in the Near East, when oil was discovered there in the early twentieth century and foreign companies obtained their first concessions. The impact and status of foreign capital in Russia had its own pattern, which was determined by a number of key factors—the country's level of economic development, social and political structures (affecting the strength and durability of political authority), and also Russia's position within the larger system of international relations.

When foreign investment did begin to flood into Russia in the late nineteenth century, it concentrated above all on the petroleum industry. While the total sum of foreign investments rose four-fold between 1894 and 1901, this infusion of outside capital amounted to a 30-fold increase for the oil industry (from 2.6 to 83.4 million rubles).[2]

This accelerated influx of foreign capital was due to a variety of factors. One set of factors was essentially economic—most notably, the high profitability of Russian enterprises and the ambition of foreign corporations to assert control over the Russian oil business. But political motives were also important; these included the escalating competition of the great powers, the attempts to dominate Russian foreign policy and the disposition of her colossal resources.

So far as the Russian government itself is concerned, its attitude toward the penetration of foreign capital was shaped by a number of factors. It is important, however, to emphasize that the regime not only looked upon foreign capital as a source for industrial investment, but also made regular application to European markets for state loans. These phenomena were closely interconnected; industrial concessions to foreigners formed an organic extension of state loan policies, and vice versa.

The interests of autocracy demanded active participation in world politics. One of the prime objectives of Witte's program was to reassert Russia's status as a world power. Its own resources, especially economic, simply did not suffice. Hence, the country needed a constant infusion of foreign loans and investments to stimulate growth in Russian industry. During Witte's tenure as minister of finance, the government relied upon a complicated mechanism to achieve this goal. Witte's system made extensive use of banks, which acted at once as an intermediary and as an agent of the Ministry of Finance. It proved possible for Witte to cast the banks in such a role because, amidst the industrial boom of the 1890s, these showed a marked increase in entrepreneurial activity.

THE ROLE OF THE ST. PETERSBURG INTERNATIONAL COMMERICAL BANK

The leading role in the foreign operations of the Ministry of Finance belonged to the St. Petersburg International Commercial Bank, which also participated actively in the organization of oil enterprises in Russia. This bank was the chief intermediary of the tsarist government in all its relations with the European stock and securities markets. Although its capital included significant amounts of foreign deposits (both German and French), in general it remained quite independent—both because it had Russian deposits and because it enjoyed various forms of state protection and support. From the late nineteenth century it was headed by A. Rothstein, a prominent businessman with extensive ties to European financial circles. It was said that Witte "does nothing whatsoever without the advice and counsel of Rothstein," who was called "the alter ego of Count Witte" and compared with the famous "Grey Eminence" who served as the key advisor of Cardinal Richelieu, Father Joseph.[3]

Concurrently with his directorship in the International Bank, Rothstein served as director of the Russo-Chinese Bank. That was no accident: until the end of the 1890s both these banks were closely associated, and in a certain sense the Russo-Chinese Bank was subordinate to the International Bank.

The period of Rothstein's leadership of the International Bank and his participation in the activities of the Ministry of Finance coincided with a strengthening in Franco-Russian political and economic relations. Rothstein was directly involved in realizing many of the initiatives of the tsarist government in its relations with France. Documents from the archive of the French Ministry of Foreign Affairs show, moreover, that the leaders in the French diplomatic apparatus were fully aware of Rothstein's role. As the French ambassador in St. Petersburg wrote to his home government, "I do not need to remind you about the confidence that he [Rothstein] enjoys with Count Witte. It seems to me that Rothstein is really the person in charge of financial enterprises in the Far East."[4]

Indeed, at times it is difficult to tell when Rothstein is acting as an entrepreneur and when he is functioning as plenipotentiary of the Ministry of Finance. Quite clearly, the interests of private capital and the tsarist government in these questions were interwoven not in some vague metaphorical sense, but in the most concrete and intimate way. It is therefore hardly surprising that official representatives of the Ministry of Finance carried out assignments from Rothstein and reported back to him.

The interrelationship between the Ministry of Finance and the International Commercial Bank (and personally between Witte and Rothstein) were based on many years of practice. But by no means was all of this intended for public consumption. Plainly, it was no accident that in the correspondence between Rothstein and the Rothschilds, Witte's name was coded as "Emil'"—a practice, significantly, that had already commenced with Witte's predecessor, I. A. Vyshnegradskii.

Later, when Witte wrote his memoirs, he referred to Rothstein as "a marvelously gifted financier and banker, an honest and intelligent person, but rather impertinent and not very likeable in his manner."[5] He remained silent, however, on the role that Rothstein played in implementing his policies as minister of finance.

It is difficult to overestimate Rothstein's influence on Witte. The French ambassador in St. Petersburg even assured his government that Rothstein inspired Witte's entire financial policies, described him as a "person of fervid imagination, very inventive, and unbelievably impertinent." He added that it was Rothstein who "carried out the major financial operations," even though it was Witte's name that subsequently became identified with them.[6] It is in fact true that, during Rothstein's tenure as head of the St. Petersburg International Commercial Bank, he did conduct more than one major financial operation. Nevertheless, the assessment by the French diplomat suffers from some exaggeration. However great Rothstein's role might have been, the chief figure was the minister of finance.

The German historian, G. Hallgarten, has called Witte "virtually a vice-tsar,"[7] and the Soviet historian M. N. Pokrovskii regarded him as the sole figure of statesmanlike dimensions in the final decades of the tsarist regime. That characterization applies fully to Witte's activities in economic and financial affairs. The minister of finance listened attentively to Rothstein's counsel and made use of his services, but always reserved the final decision to himself.

While Witte did pursue a policy of attracting foreign capital to Russia, he did so within the context of a complex system of state supervision. One of his primary goals in controlling the activity of oil companies was to prevent Standard Oil—the chief rival of Russian oil on foreign markets—from gaining entry to Russia.

Nevertheless, and notwithstanding the opposition of the class of noble landowners (who feared that the penetration of foreign capital and development of industry would undermine their own position), Witte did all in his power to facilitate the access of French and English capital to the Russian oil business.

To be sure, German banks had expressed a serious interest, but it was French and English capital that won the privileged position. The general tendencies in Russia's financial relations, which were shifting from the German to the French stock and security markets, worked against the plans of German capital and clearly promoted the French and English projects.

THE CONSOLIDATION OF THE POSITION OF THE FRENCH ROTHSCHILDS

Soon after the Rothschilds established themselves in Baku, they petitioned the tsarist government for permission to increase the capital of their Caspian and Black Sea Company.[8] The request was submitted in 1887, but because it did not immediately obtain approval, the Parisian bankers decided to give the company a "loan" in excess of 3 million rubles (which was more than twice the amount of the fixed capital—1.5 million rubles—that had been approved for incorporation by the tsarist government).[9] The loan provided the means to create a monetary fund, which was then used to finance a large number of Baku firms. The dearth of credit constantly experienced by the oil industry in Baku contributed to the realization of the Rothschilds' plans. As already pointed out, they succeeded in creating a bloc of satellite firms around their Caspian and Black Sea Company, all linked by their financial dependence on the Rothschilds' company.

In the 1890s the Rothschilds' Banking House significantly expanded its role in the Russian petroleum industry—obviously in an effort to exploit the development of Franco-Russian financial ties following the establishment of a new political alliance between the two countries in 1891.[10] As a Russian businessman residing in Paris observed, "people interested in Russian affairs have a great influence on the Paris stock and securities exchange."[11] By the mid-1890s the "passion" for Russian stocks had somewhat abated, for the French market had been saturated with Russian securities.[12]

Alphonse de Rothschild even noted that "for some time" a certain "opposition" was apparent "with respect to Russian finances."[13] Nevertheless, such sentiments had no effect upon the zeal for investment in the oil business.

The Russian industrial boom that followed the economic crisis of 1893 gave a new impulse to the efforts of the French financiers. With Skal'kovskii's assistance, at the end of 1895 the Rothschilds finally obtained the long-awaited permission to increase the fixed capital of the Caspian and Black Sea Company. That of course did not mark the end of Skal'kovskii's services, but henceforth the chief canal for the French proved to be the St. Petersburg International Commercial Bank. It was thanks to this bank that the Rothschilds—making use of their position as creditors of the tsarist government—acquired the ability to put pressure on the Ministry of Finance and on Witte personally. Striking evidence of their power is to be found in the confidential correspondence of J. Aron and A. Rothstein (in the archive of the St. Petersburg International Bank), which provides an extraordinarily rich picture of the machinations by financial capital.

From the mid-1890s business circles in France, with the active support of the government, more and more frequently endeavored to exploit Russia's dependence on French capital. That obtained even with respect to the sensitive issue of the plight of Russian Jews. Over the many years that the Rothschilds negotiated with the tsarist government they invariably expressed their dissatisfaction with the status of Jews in the Russian Empire. They demanded that Jews have the same rights of enterpreneurship enjoyed by the rest of the population, for existing tsarist law limited their rights (similar to foreigners) to participate in business enterprises. Hence, the Jewish directors of the Rothschilds' Baku firm were in an inferior position to their main competitor, the Nobel company. On the other hand, the whole question of discrimination against Jews (and, subsequently, their persecution after the pogroms of the early 1900s) constituted a serious complication in the negotiations and relations of tsarist authorities with Jewish financiers in France, England, and other countries. It must be said, however, that for the sake of concluding an advantageous deal, the Rothschilds were willing to close their eyes to this whole problem. That is what happened in 1892, when they attempted to obtain an advantageous arrangement for themselves in Russia. As a result of the agreement, they expected to be on equal terms with Nobel in Baku and impelled the tsarist government to put pressure on the French authorities to cut the tariffs on oil, thereby killing two birds with one stone. As Hallgarten has observed: "While the liberal press of the entire world (well informed from certain quarters in Paris) raised a commotion about the maltreatment of Jews in Russia, the Rothschilds in total secrecy conducted negotiations with the Russians about the conditions for a Franco-Russian trade agreement, according to which Russia was supposed to compel the French to reduce the customs on oil." Hence, the Rothschilds had "thereby put a foreign government on a leash in order (with this commercial treaty) to force its fatherland to make concessions in the interest of its production outside its own country."[15] For the Rothschilds this had a special meaning, given the French petroleum market, which was under the control of an association of oil refineries. France imported crude oil from abroad and controlled the distillation process itself; as a result, she did not become the arena for "kerosene wars," as was the case in Germany, for example. At any rate, a reduction in tariffs on the import of crude oil into France would simply give the Rothschilds an opportunity to extract additional profits from the domestic oil business.

To resolve all important questions pertaining to Russia, the Rothschilds' representatives acted through Rothstein, but in particularly complicated cases, they appealed directly to Witte. For instance, in the second half of 1896 Aron was forced to visit the tsarist Minister of Finance regarding attacks in the Russian press and the conflict over the tariff on the kerosene that the Rothschilds exported from Russia. A small official conference was convoked to discuss these issues. According to Aron, Witte "began with a sharply worded speech" against the insinuations in the press and ordered that "an official repudiation" be issued. The minister's position "was correct in every respect." "He wants to throw a bone to the customs officials' dog. But we must defend ourselves and arrange things so that this bone is as small as possible." After the conference Aron remained to dine alone with the minister and had a "private conversation" with him. Witte expressed "his regrets over the fact that, because of his relations with Mr. Rothschild, he could not put a stop to this matter." He emphasized that, even as it was, he was subject to attacks on the grounds that the Ministry of Finance allegedly "is inspired in all its activities by Mr. Rothstein, who in turn is spokesman for the interests of B[aron] Alphonse R[othschild]." During this conversation Aron also raised some "other questions" and was, in his own words, "delighted by the answers I received, because they conform to our views."[16]

The Rothschilds' firm assessed perfectly the significance of the minister's position and, as Aron candidly declared, that was precisely the basis for hopes to realize their oil projects in Russia. Noting that "real insanity is raging" on the European stock exchanges "with regard to everything associated with the Russian oil business," Aron portrayed things as though the Rothschilds were dispassionate and composed in seeking to achieve their goals.[17] However, the French financiers in fact did reveal clear impatience in wishing to achieve a further expansion of their oil holdings in Baku and to consolidate their interests in Russia. In the mid-1890s the Rothschilds made vigorous efforts to establish themselves on the domestic Russian market. Aron himself came twice to Russia for this very purpose. At the end of 1896, after a brief stop in St. Petersburg (where the International Bank gave him "a cordial reception"), the Rothschilds' representative then journeyed to Baku, where, in his own words, "I became so intensely preoccupied with business that I neglected my family and friends." "I maintain contact with Paris only by means of telegrams—I have absolutely no time to take up the pen and write a few letters."

In 1893-1896 the Russian oil business experienced such an explosive growth that the French businessman was utterly astonished by the colossal possibilities that he saw for industry and trade in Baku. "In the three years since my last visit," he wrote, "the oil industry has undergone enormous development, and I believe that, ... because of the rising consumption of liquid fuel in those areas of Russia linked through transportation by the Volga, we will come to realize new progress in the future."[18]

It was not the export trade (where the Rothschilds already had a strong position), but trade on the domestic market—"in the regions of Russia, linked by transport with the Volga"—that now stood at the center of attention on Rue Laffitte, the Parisian street where the Rothschild bank had its offices. It was necessary to act with dispatch, all the more since their main rival, Nobel, already had at its disposal a well-developed commercial network inside Russia for the distribution of fuel oil.

The first firm to attract the attention of the French financiers was the Baku company of G. Tagieff, which at the time was experiencing serious financial difficulties. The Parisian bankers had already begun to look the firm over in early 1896. Besides consulting with Skal'kovskii about the possible terms and expediency of such a purchase, the Rothschilds also called upon the services of Rothstein to help prepare a concrete plan of action. That plan proposed not only to acquire the Tagieff enterprise, but also to use it as the basis for a commercial organization—the Baku Merchant Bank. The Tagieff enterprise seemed ideally suited for this goal. As the director of the Rothschilds' Caspian and Black Sea Company (A. Feigl) pointed out, "the firm has long been active in discounting commercial acceptances, rentals, etc., etc., and also making advances on goods being sent into internal areas of the country, so that today the firm already has a quite significant clientele, which, of course, could be significantly expanded." Moreover, the architects of this project deemed it essential to use Tagieff's personal connections and emphasized that the Baku entrepreneur enjoys "a very great influence" and "the foundation of such a banking enterprise with the Tagieff firm at its head would only lead to good success."[19] The aim of obtaining the Tagieff firm—one of the most influential business houses in Baku—was to transform it into a joint-stock company, which would become the backbone of the Rothschilds' activities in the Russian oil business. But the negotiations with Tagieff collapsed and nothing came of the Rothschilds' elaborate plans.

The Rothschilds then shifted their attention to another independent enterprise—the commercial firm of Poliak and Sons. It was active in shipping oil along the Volga, owned storage tanks at wharves along the Volga, and had ships for the bulk transport of oil. The Rothschilds decided to make this firm the base for a company called Mazut (literally, "Residual Fuel Oil"), which would "retain the character of a Russian enterprise" in order to avoid difficulties in obtaining government approval of its charter. The Rothschilds would, however, hold the controlling stock, a fact that was not intended for public knowledge.[20]

In a letter to Rothstein, Baron Alphonse de Rothschild expressed his confidence that Mazut would have great chances of success and that the new enterprise will "yield a profit on the capital that has been invested."[21] This letter came literally a few days after Aron had asked Rothstein to "prepare a draft charter" and to obtain Witte's permission. "As soon as the charter is prepared," wrote Aron, "it must be presented to the minister of finance, and I think that no one can do this better than you."[22]

The first attempt to obtain the approval of the Ministry of Finance, however, ended in failure: the government rejected the Rothschilds' application, with the explanation that foreigners were forbidden to participate in shipping on the Caspian Sea. But that did not cause Paris to give up. After a conversation with Baron Alphonse de Rothschild, Aron once again applied to Rothstein for assistance, expressing the hope that, thanks to his "influence in high places" and "Emil's desire to be agreeable to my firm," the plan to create Mazut could nevertheless be achieved.[23] Though the Rothschilds were evidently becoming impatient, "Emil'"-Witte nevertheless did not make haste.

To consider this stalemate in Russia, the Rothschilds convoked a special meeting in Paris in August 1897, which was attended by a representative of the St. Petersburg International Commercial Bank (A. Davidov). In summing up the results of this meeting, Aron observed: "We came to an agreement regarding the means to be taken in exerting their influence on the minister [of finance]."[24] According to Davidov, these "means" consisted in continuing to put pressure on Witte. The Rothschilds, he wrote, "count" on the "support" of Rothstein and on Witte, and expressed the hope that "Emil' will make up his mind" to submit for government approval the charter of Mazut "just as it is," including the article on its right to engage in shipping on the Caspian Sea (which they "have absolutely no intention of renouncing").[25]

Rothstein communicated all this in a special meeting with Witte. But the latter still avoided a definitive answer and merely inquired whether the Rothschilds might not be better advised to build an oil pipeline between the source of the Volga and Moscow. Although the Minister of Finance did not give a flat refusal, he sought to procrastinate, evidently in order to make the French financiers more forthcoming in the impending negotiations for new loans that the tsarist government so sorely needed.

In the spring of 1897 Russia finally succeeded in obtaining a loan on the Berlin stock exchange, the first time after a considerable interval. This agreement came against a background of political statements in the press and the Reichstag regarding the desirability of restoring "friendship" with Russia. Naturally, none of this could have pleased France. Under the terms of the Franco-Russian alliance, the "financial influence of Germany" loomed as a clear menace—such was the view at the Quai d'Orsay.[26] News that Russia had arranged a loan with Germany produced a similarly negative response at the Rothschilds' headquarters on Rue Laffitte. Baron Alphonse de Rothschild was indignant over Witte's "hastiness," and testily complained to Rothstein that "the minister used two different measures in his actions": he applied "the greatest severity" toward the French, but accorded the Germans "every kind of advantage."[27]

Indignation in Paris rose a notch higher when Witte began to explore ways to tap the British money market and, in a transparent effort to mollify British capital, offered the British an oil concession in Baku. In the end, the English bought the very same firm of Tagieff that had first interested the Rothschilds. "According to our information," fumed Aron, "the English acquired the enterprise only after they had received a firm assurance from the Russian government that no kind of obstacles would be put in the way of their running the business." Vexation bristled in every line of the letter from the Rothschilds' representative, who asked: "How can the law that they use against us be compatible with such assurances?" And "what about the Nobel firm, which possesses a flotilla on the Caspian Sea?"[28] Why did they deny the Rothschilds what was already owned by their main competitor in the Russian oil business? "You know better than I," Aron reminded Rothstein, "that a significant quantity of stock in this company [of Nobel] are in the hands of foreigners, but the restrictive article [in the legislation] has in mind not only the nature of the securities, but also the character of those holding the stock—who, without any exceptions, must be Russians."[29]

Why then, asked the Rothschilds as the tsarist government reviewed their application for Mazut, did it treat the Nobel firm so favorably and make concessions to British capital? The Parisian entrepreneurs therefore requested Rothstein "to have a completely clear idea about these two cases" and demand analogous rights for the Rothschilds. Nevertheless, Paris understood what Witte wanted—concessions with respect to loans. Although the fact that a Russian loan had been floated in Germany provoked the

Rothschilds' unconcealed irritation, the latter tried to clarify Witte's "general intentions" and expressed a willingness "to be nice to him" in order to "strengthen his disposition toward us." Thereupon Baron Alphonse, in order to "meet the minister's wish," reported his decision to participate in placing the next Russian loan on the Paris stock and securities exchange.[30]

Shortly after receiving this news, Witte approved the charter for Mazut and, in compliance with the legal formalities, obtained the requisite confirmation of the Committee of Ministers. Although the Rothschilds' new company was formally constituted as a Russian enterprise, that it most certainly was not—either in terms of the stocks ownership or the composition of its board of directors. Of the 16,000 shares of stock, the Rothschilds owned 9600, leaving 4000 for Poliak and 2400 for the International Commercial Bank. The latter distributed its shares among its traditional partners (as co-participants): the Russian Bank for Foreign Trade, the St. Petersburg Discount and Loan Bank, the Hamburg Bank of Max Warburg and others. According to the terms of the agreement, Rothstein and his partners were obliged "for five years to put the stocks in the syndicate at the disposition of the Rothschilds" and to be "in agreement with all the decisions which might be taken by the Paris house [of the Rothschilds]."[31] Of the five places on the board of directors, three belonged to the Rothschilds and one each to Poliak and Rothstein. Upon the initiative of Baron Alphonse, Rothstein was elected chairman of the council. In return, the board admitted a Rothschild business partner, G. Spitzer, who was assigned a large bloc of stocks in the International Commercial Bank. This operation marked the consolidation of Russian and French financial interests and accompanied steps being taken along the path of further political rapprochement. Although there was no direct connection between the two phenomena, they formed part of a common internal logic.

THE INFLUX OF BRITISH CAPITAL

The significance of the petroleum trade, and later of fuel as a strategic material had long been understood in the business and political circles of England. Therefore by the turn of the century, when various foreign companies entered the competition for

Russian oil, British capital was fully prepared to join the fray. Not only the Samuel, Samuel's Co., which had become actively involved in the kerosene trade in the East, but other English firms as well displayed an interest in the Russian oil business, on the conviction that it would bring a good profit.

It was not, however, simply a matter of the profitability of the kerosene trade, for the oil question had aroused the interest of English political circles from the very first. That interest derived from the fact that England, even before most other countries, understood oil's potential use as a fuel for military naval vessels to make the fleet much more mobile. The idea of converting the navy to oil-burning engines was first broached in the 1880s by Lord J. Fischer, who, two decades later became lord of the admiralty and worked to win parliamentary approval for this measure. Much time would pass before the "oil maniac" Fischer would see his efforts realized. But awareness of oil's strategic significance for the navy existed in the highest British circles long before parliament approved steps to renovate the fuel-base of the nation's fleet.

Vested interests therefore sought to strengthen England's position by acquiring oil fields. In the 1890s the English established companies to explore and drill for oil in the Far East—in Burma and on the island of Borneo in the Dutch Indies. But the results were disappointing. By the start of the twentieth century, Burma produced only 1 percent of the world's total output, and that in the Dutch Indies amounted to only 1.5 percent (of which only a small part was the share accorded to England). Under these circumstances, British capital took a keen interest in the Russian oil industry that, precisely in these years, was experiencing a boom.

The end of the nineteenth century merely signals the beginning of England's acquisition of oil fields, but it had already made considerable headway in commercial dealings with petroleum products. That is hardly surprising, for the "mistress of the seas" had the most powerful merchant marine in the world. Many ports to the east of the Suez Canal were linked with London by continual shipping lines, and English agencies to sell kerosene sprang up like mushrooms and served as outposts of English expansion. By the close of the 1890s, the Samuel and Samuel Co. alone owned 30 ocean-going steam-powered oil tankers and dozens of oil storage facilities in various ports. By 1900 the total number of its agencies amounted to some 320 firms, located in every port throughout the East, in African colonies, British India, Siam, China, Japan, the Dutch Indies, Australia, Zanzibar, the Near and Middle East.

An important landmark for British capital came in 1897 with the creation of a powerful shipping company for the oil trade—"the Shell Transport and Trading Co.," which was established through the amalgamation of small firms interested in the oil trade around the Samuel, Samuel's Co. In addition to this new group, there were also a number of other companies, the largest of which—Suart and Gladstone—led the opposition against the newly formed firm of Shell. Nevertheless, establishment of Shell marked an important step toward the consolidation of British petroleum interests.

After the turn of the century, the Shell Company re-outfitted its steamships to use petroleum fuel. Samuel urged the British admiralty to follow suit and expressed a willingness to supply the fleet with the necessary fuel. That agitation of course had its weak point: at the time England still lacked its own oil production and, practically speaking, was totally dependent upon the delivery of Russian oil.

To solve this problem, British capital rushed to gain a foothold in Russian oil production. Initiative came from rivals of Samuel—A. Suart and G. Gladstone. In 1896 Suart established the first English oil producing firm in Baku—the "European Oil Co.," with a capital of 5 million rubles. The formation of Shell caused Samuel's rivals to take more energetic measures. At the end of 1897 the Gladstone Co. purchased the oil fields of Tagieff, which it thus snatched from Samuel's allies, the Parisian Rothschilds, who had spent a whole year trying to persuade the Baku industrialist to sell out. The English outplayed the French and reorganized the former Tagieff Co. into a large British oil-producing firm, which had a capital of 12 million rubles and became known in business papers, the press, and literature as "Oleum." To secure the success of the enterprise, the director of the English Bank, E. Hubbard (who owned textile plants near Petersburg and had a good knowledge of Russia) was appointed to head the company.

Samuel emulated the example of Suart and Gladstone: almost simultaneous with the purchase of Tagieff, his contractors bought out the firm of "Shibaeff [Shibaev] & Co.," which became the basis of the "Shibaeff Petroleum Co.," with a capital of 6.5 million rubles.

British enterprises in Russia proved exceedingly profitable. The greatest success belong to Oleum, whose oil fields yielded powerful geysers of oil. Chambers, the American consul in Batumi, observed that after the very first year English operations had already recouped the purchase money and also turned a significant profit. Their astonishing success provoked a sensation on the London stock market. Chambers even suggested that in the future the entire Baku business could fall into the hands of British entrepreneurs.[32]

Business was also turning out well in the enterprises owned by the group of Lane and Samuel. During the first five years, net profits of the former Shibaeff company completely paid off the expenditures for its purchase. The income that the English owners received each year from their oil enterprises in Russia significantly exceeded the profit from their investments in England. The rapid growth of the petroleum industry during these years and, especially, the rise in demand for fuel oil opened the possibility of still higher profits. As a result, Baku was soon inundated by throngs of English businessmen in search of an easy windfall. "Foreigners are to be seen everywhere," wrote the American consul, "so much so that Baku will soon be transformed into an English town. The investments are considered good, and if no objections are raised, no one will be surprised if in the near future the whole naphtha trade gets into the hands of the English."[33]

Without first securing permission from tsarist authorities, English businessmen and stock-jobbers discussed the prospects for entrepreneurial activity in Russia—much to the indignation of G. Vilenkin, the Russian finance ministry's agent in England. He proposed to put an article in English newspapers discussing the rules for the admission of foreigners to the Russian oil business and thus put the impudent operators in their place.[34]

However, the very thing that Vilenkin proposed to interdict was precisely what the Ministry of Finance decided to encourage. Even before the purchase of the Tagieff business, English entrepreneurs had received a "firm assurance" from the Russian ambassador in London (E. E. Staal') that the tsarist government will not pose "any kind of obstacles" for British companies interested in exploiting the oil-fields of Russia. The evidence for this, preserved in the extant correspondence of the Rothschilds,[35] may seem incredible, insofar as it is known that the current Minister of Foreign Affairs (M. N. Murav'ev) was one of the most active opponents of admitting foreign capital into Russia. Nevertheless, cognizant of Staal's anglophile sentiments, Witte decided to win him over. As documents in the London embassy show, the Ministry of Finance regularly informed Staal' of its intentions in an effort to solicit his support.[36] There can hardly be any doubts that the ambassador's clarification of the rules for admitting foreign companies to Russia, perceived as a "firm assurance" that the English will not encounter obstacles, was made not only with the permission, but even at the behest of Witte. Vilenkin's predecessor as the Finance Ministry's agent in London, G. Kamenskii, verbally and in the writing repeatedly confirmed the promise given the English.[37] In January 1898, during a reception at the British embassy in Petersburg, Witte finally succeeded in persuading the British envoy (E. Hubbard) of his "favorable attention" and, subsequently, reaffirmed this attitude to the English businessman, with a promise to give their enterprises the necessary assistance.[38]

Witte decided to strengthen his representation in London and, with this goal in mind, replaced the novice Vilenkin with the more respectable S. S. Tatishchev, who had many years' experience in diplomatic service.[39] Upon his arrival in London, Witte's new plenipotentiary immediately established relations with the largest banks, above all with the English Rothschild, to whom he declared that he wishes to be "just as accredited" before Rothschild "as Staal' is accredited with Queen Victoria." For his part, the English banker expressed satisfaction with Tatishchev's appointment, thanks to which he counted on maintaining direct relations with Witte through his "permanent representative," "the Russian gentleman," toward whom he has "greater trust" and "speaks his mind with greater candor than with transitional and accidental emissaries." But that was as far as the conversation went. On the question of loans, Rothschild said that, until Anglo-Russian differences were resolved and until "dark clouds disappear from the political horizon," Russia can hardly count on the London monetary market. To be sure, Tatishchev tried to entice him with the economic possibilities that would be opened in Russia for English capital, as he reported on recent reductions of import duties for goods "chiefly of English manufacture" and emphasized that British entrepreneurs were assured "enormous profits, unknown anywhere else in Western Europe."[40]

In fact, whereas investors could expect only a 3 to 5 percent return on their capital investments in England, they were assured 10 to 20 percent in Russia.[41] Those facts must surely have been known to the Rothschilds. But they remained deaf to Tatishchev's words; yet even tantalizing allusions to "the absolutely unbelievable dividends of 40, 50 and 75 percent" seemingly failed to have much effect.

Meanwhile, Witte did not abandon hope of achieving his goal by making concessions to English capital and arousing its interest in Russian trade and industry—which was all within his broader economic program.

But in Russia itself, within ruling circles, Witte's policy encountered opposition from conservative gentry landowners, who did not hesitate to give vent to their discontent. Among those who opposed admitting English capital to Russian oil was the tsar's brother-in-law, Grand Duke Aleksandr Mikhailovich, who later became an active member of the Bezobrazov circle that helped inspire Russia's adventurist policies in the Far East. The Grand Duke sent Nicholas a letter and a special memorandum, which, in all likelihood, had been composed by the conservative journalist S. Sharapov, who was then conducting a press campaign against Witte's policy of soliciting foreign capital.[42]

As the pretext for writing the tsar, the Grand Duke used the attempt by Hubbard and Co. to expand their recently acquired holdings in Baku by adding some state properties that were contiguous with the Tagieff oil fields. Indeed, before the English had had time to digest their first acquisition, they advanced new claims, which the Grand Duke described as an attempt to "seize control of the oil industry." He also noted that, if the government acceded to their request, this would be "equivalent to transferring the entire petroleum industry to the exclusive utilization of the English and equivalent to taking the entire petroleum industry out of Russian hands."[43] However, the main motive behind the Grand Duke's "patriotism" was his desire to seize for himself a tasty morsel that the English wanted—so that, as was usually the case, he could later resell the property to the highest bidder (not excluding the English entrepreneurs). To defend the national interests of Russia, the Grand Duke proposed that he, not the English, be consigned these oil fields.[44] In a word, the lofty arguments of this representative from the imperial family were simply a cover for a base attempt at personal gain.

Although Nicholas II was favorably disposed toward the faction of conservative gentry landowners (whose outlook the Grand Duke represented), circumstances did not work in their favor. On May 19/31, 1898, when the gentry party attempted to have the Committee of Ministers adopt their proposals to restrict foreign landholding in the Caucasus, Witte gave battle to his adversaries. He began by warning of the "serious dangers" that would ensue from "the undesirable impression" that publication of the new rules would make "on industrial and commercial spheres in Western Europe" and especially in England. Witte also tried to tempt his opponents' self-interest by noting the possible dangers that the English market could harbor for the sale of Russian agricultural products. Then, after describing the critical condition of Russian state credit in France, he drew attention to the significance of the monetary market. "The above considerations," he declared, "have forced the Ministry of Finance to give special attention to the conditions of the English market, which, furthermore, is more extensive than that in France." Witte urged that no new limits be imposed and, on the contrary, proposed to dispel the uncertainty prevalent in English business circles and especially "the lack of trust in the stability of rules determining the status of foreign industrialists and businessmen in Russia."[45] Turning next to the question of permitting English firms to acquire oil fields in Baku, Witte insisted that no obstacles be raised. For that purpose, he reported, the Ministry of Finance had already placed a special announcement in English newspapers about the rules "regarding this industrial branch and the participation of foreigners."[46] This alone had sufficed to have "English capital flow into the oil business," but then the dissemination of rumors that these deals would not be confirmed had caused English capitalists to come to St. Petersburg "for fuller ... explanations."[47] Witte did not say that he had met with them, vowed his support, and assured them of a favorable outcome. He then demanded that the Committee of Ministers resolve the matter on his terms, and repeated his admonition that the introduction of new rules "will alarm foreign spheres" and that Russia will suffer "losses on the international market that will be difficult to make up."[48]

Witte was certainly in a difficult position: at the very time the Committee of Ministers was discussing this question, Prince G. S. Golitsyn (the chief civil official in the Caucasus) sent the emperor a memorandum urging tight new restrictions on foreign capital, and Nicholas himself wrote the following resolution on the memorandum: "I also find these measures necessary."[49] Nevertheless, Witte succeeded in causing the ranks of his foes to waver. Under the weight of Witte's arguments, even Golitsyn felt obliged to retract his categorical statements. Although Witte did not succeed in achieving complete success at this first discussion, he did dissuade the Committee from publishing its resolution (which, in any case, softened the language of the original text) and to exclude "all decrees in any way infringing the rights already acquired by foreigners."[50]

Two weeks later, on June 3/15, 1898, at a special conference summoned to deliberate the memorandum of Grand Duke Aleksandr Mikhailovich about the participation of English capitalists in the Baku oil industry, Witte won a decisive victory. The conference supported the policy of the minister of finance, a decision that subsequently endorsed by the tsar.[51] Thereafter Witte had no difficulty persuading the Committee of Ministers to confirm the charters of the English oil companies, which thus obtained the final sanction required by Russian law.

That is how British capital legalized its position in the Russian oil business. But neither Suart and Gladstone, nor the representatives of Shell, intended to stop with what they had thus far achieved. To compensate Witte for his efforts, English banks undertook to float the bonds with a 4-percent Russian annuity, which a year earlier the French had declined to do.[52] In June 1898, during the very same days that Petersburg decided to admit English companies to the Russian petroleum business, these bonds were admitted for quotation on the London stock exchange. Now the English could count on new signs of attention and favor from the tsarist government.

English companies were all the more in need of Witte's support, for the foes of foreign capital, despite their defeat at the Special Conference in 1898, had no intention of throwing down their arms. The conservative press, and above all *Russkii trud* (which was edited by S. Sharapov and oriented toward gentry landowner interests) and also the newspaper *Moskovskie vedomosti,* launched a crusade against the admission of foreign capital, which would accelerate the country's capitalist development and strengthen the position of the bourgeoisie, thereby relegating gentry landowners into a subordinate role.[53] *Moskovskie vedomosti* did not mince words when it complained that the policy of attracting foreign capital will transform Russia into "an exclusively industrial state."[54] Opposition also came from certain industrial circles, in particular the Moscow bourgeoisie, whose enterprises were dependent upon oil deliveries from Baku. At the end of 1898 the largest Moscow industrialists appealed to the Committee of the Moscow Stock Exchange, complaining about a sharp rise in oil prices which they ascribed to the activity of foreign companies in Baku. The Committee endorsed their view and lodged a protest of its own.[55] To be sure, these declarations were countered by a barrage of rebuttals from supporters of admitting foreign capital, who depicted the "gains" from foreign capital in the most glowing terms and praised Witte's policies. This theme figured in many contributions to the publications of the Ministry of Finance and also to *Torgovo-promyshlennaia gazeta;* the same subject was raised in articles by the K. A. Skal'kovskii and the liberal professor I. I. Ianzhul' and even provoked an entire "study" by B. F. Brandt.[56]

Despite this defense of Witte's position, his foes were able to exploit the connections of a court camarilla, taking advantage of Grand Duke Aleksandr Mikhailovich's influence, in late 1898 and early 1899 induced the tsar to begin changing his position.[57] Nicholas II would have willingly reversed himself fully were it not for the arguments of his Minister of Finance. As before, Witte emphasized that the government needed new loans abroad (which seemed indispensable for the survival of tsarism), and also argued that attracting foreign capital is a necessary (if not the sole) method for promoting the country's economic development. In February 1899 Witte submitted to the tsar an extensive memorandum ("On the Program of the Commercial-Industrial Policy of the Empire") and a month later succeeded in having this document endorsed by a specially convoked meeting of the Council of Ministers presided over by the tsar himself.[58]

After this Witte had no particular difficulty in prevailing over his adversaries in a new Special Conference, summoned in April to consider the question of whether to permit English companies to participate in the next auctioning of lands containing oil reserves. It is highly indicative that, whereas M. N. Murav'ev (the minister of foreign affairs) had figured as one of Witte's chief opponents at the conference in March, he now changed his position. Responding to changes in the tsar's views, Murav'ev prepared a special memorandum for the meeting of March 17, which he read at a session of the Council of Ministers. Raising no objections to foreign capital "in the form of cheap bond loans," the minister of foreign affairs voiced his opposition to admitting capital "in the form of joint-stock enterprises." He warned that "along with [the enterprises] will come ideals and strivings endemic to the capitalist order, [which will] penetrate the general population, and groups of foreigners—with the assistance of trusts and syndicates—will take control of the country's natural resources and the real owners of large economic units."[59] After Witte gained the upper hand at this meeting and won the tsar's support, Murav'ev hastily changed his position. At a new conference he spoke immediately after Witte and affirmed that "he is in total agreement with the minister of finance" and supports his proposals completely, since "the closer the financial and economic ties between states, the better and stronger their political relations."[60] This was a 180 degree turn. The new conference supported the decision of 1898, decreeing that it recognizes "the participation of foreigners and foreign companies in the Russian oil industry as useful and desirable."[61]

The applications of English firms to participate in the next auctions of oil-bearing lands were approved. According to the calculations of the Minister of Finance, London should show appreciation for his efforts. So as not to leave any doubts on this score, a "secret report" was leaked to the English press containing Witte's speech at the session of the Committee of Ministers in May 1898, where he was especially zealous in supporting the admission of English capital to Russia. Published on April 26, 1899 in the *Times,* this report created a sensation, provoking a panic in French financial circles—whose services Witte intended to spurn in favor of a new orientation toward London.[62] Even officials in the tsarist Ministry of Foreign Affairs were puzzled by the publication in the *Times.* Well-informed people, however, had no doubt that this was Witte's handiwork. For example, G. Spitzer (a representative of the Parisian Rothschilds), immediately suspected Witte's involvement in the *Times* publication: "In my judgment," he wrote, "this so-called unauthorized divulging of a state secret could very well have been done with the permission of Emil'."[63]

In this case, what were Witte's goals and how far-reaching were they? Did he contemplate a radical change in his relations with Paris, or was this simply a clever maneuver? Witte could not, of course, be serious about declining the services of the French stock exchange. Notwithstanding all the limitations on Russian loans in France, only the French market was capable of satisfying tsarism's insatiable demand for money. This fact, without doubt, continued to form one of the motives of the Franco-Russian alliance, which, in an exchange of diplomatic notes, both sides reaffirmed a few months later. Witte of course could not have planned to undermine an alliance that formed the cornerstone of Russian foreign policy. At the same time, in turning to London, the Minister of Finance not only took into account the realities of the English money market, but also the possible improvement in Anglo-Russian relations more generally.

The fact is that, during the nearly six months of deliberations about admitting English capital to the Russian oil industry, attempts were simultaneously being made to resolve problems in Anglo-Russian political relations. In January 1898, Prime Minister R. Salisbury charged the English ambassador in Petersburg, N. O'Connor, "if practicable," to ask Witte—whom London, like other European capitals, regarded as the most adroit and influential member of the cabinet—whether it is possible that England and Russia should work together." In doing so Salisbury had in mind a coordinated policy in China, with a promise "to further Russian commercial interests."[64]

In July 1898, one month after Russian securities had been placed on the London stock exchange, O'Connor had a lengthy discussion with Witte and reported that "I have received many assurances of his desire for a cordial political understanding with England." In communicating this news to Salisbury, the ambassador emphasized that "I believe his assurances are no less sincere that they are mainly founded upon the necessity of gaining over the London money market. He wants British capital invested in the Russian state bonds and in Russian industrial enterprises." O'Connor added that he "could always count upon his [Witte's] being a strong advocate of peace and of a cordial and close understanding with Great Britain."[65]

The negotiations with Witte had begun still earlier, but this was the departing visit of the ambassador when he completed his service in St. Petersburg in 1895-1898. "During I may say the whole period of my residence here," O'Connor wrote, "I have had the advantage of particularly intimate relations with Monsier Witte and I have been often much indebted to him for honest and clear views in regard to various political questions." In reflecting on the Witte's motives, O'Connor suggested that his interlocutor had doubts about the French money markets over the long term and volunteered his belief that "the country at large is behind him in a strong desire for the maintenance of peace and for avoidance of the French market as less responsible or less reliable in the long run." In his view Witte's position was to be attributed "in part to a growing jealousy of the proceedings of Germany and her quasi-commercial, quasi-political activity at Constantinople and Asia Minor." The suggestions of the British diplomat cannot be said to lack perspicacity; some were soon to be confirmed completely. There were also some grounds for O'Connor's conclusion that "the key note" in the finance minister's policy was the desire to ensure that "the political program" is "in harmony with Russian financial necessities."[66] To be sure, the negotiations that began in 1898 by the ambassador in St. Petersburg were completed in April 1899, but the result was far from providing a general resolution to the problems in Anglo-Russian relations. An understanding was reached only with respect to a private question—the spheres of railway construction in China. But both sides regarded this agreement as a success that opened the door to a further dialogue.[67]

Witte in fact did regard the problem in the broad context of international relations and financial ties. For the sake of achieving this goal, he was ready to make concessions and reciprocal measures, as in the case of admitting English capital to the Russian oil industry and thereby giving British entrepreneurs an entirely special position. But realization of this policy encountered resistance from influential groups within Russia and made the unresolved questions in Anglo-Russian relations still more complicated.

In early 1899, as a scheme directly related to Russian loan policies and to efforts to increase British holdings in Baku, a proposal was worked out to create an Anglo-Russian bank. Almost simultaneously, the Suart-Gladstone group and the representatives of Shell submitted applications to expand their share in the Baku oil business, which, if successful, would raise them to the level of the biggest magnates in Russian oil. Both groups laid claims to large tracts of state oil-bearing lands and the Apsheron Peninsula and proposed to establish a semi-governmental company, backed by English capital. At issue was the most promising oil fields in the area of Bibi-Eibat.

Witte transmitted the proposals (sent by Tatishchev in London) for review to the Minister of State Domains, A. S. Ermolov. The latter immediately voiced his opposition to the schemes, observing that the English "make it their aim to receive for exploitation the best land tracts of Bibi-Eibat area."[68] True, after the English representatives themselves had arrived in St. Petersburg and began negotiations with Ermolov, he took a somewhat more moderate position. But in the final analysis both Ermolov and Witte decided to reject the proposal, having come to the conclusion that, once the government granted the English this concession, it would receive nothing from this in return. As a result, the special conference convoked to review this proposal found it unacceptable.[69]

Notwithstanding this decisive rejection, the English businessmen still hoped to realize their plans. Particularly active was Lane, who bombarded the St. Petersburg International Commercial Bank with letters.

It was precisely at this point that Rothstein advanced the idea of creating an Anglo-Russian Bank. His proposal took into account the interest of British capital in the Russian oil business, but its main goal lay elsewhere: Rothstein wanted to channel the entire British adventure in such a way as to organize something like a bank of industrial credit (with a branch in London) that could be used to float the stocks and bonds of Russian enterprises on the London stock exchange. The idea was by no means new. At the end of 1896 an analogous Franco-Russian financial enterprise (Société Générale pour l'Industrie en Russie) was created with the active participation of the St. Petersburg International Commercial Bank. It combined "the most powerful forces of the French financial world," but survived only two years, suspending operations in early 1899.[70]

In proposing to form the Anglo-Russian Bank, Rothstein was guided by approximately the same ideas that led to the creation of the Société-Générale. He expected to earn a big profit during the founding operations, while keeping initiative and control of the new institution's activity in his own hands. But it was precisely this element that the English found objectionable, as Lane informed Rothstein upon receiving the draft proposal for the Anglo-Russian Bank. Suart and Gladstone, for their part, did not want to have anything to do with the International Commercial Bank, but created their own independent enterprise (Eastern Oil Co.), which announced a broad program of investment in the economy of tsarist Russia. This scheme foundered, however, for it failed to secure the sanction of the Ministry of Finance.

Meanwhile, Lane and other English capitalists, having drafted their own plan to create an Anglo-Russian Bank, appealed to Witte and almost received his support. Participating in the negotiation was the influential British financier, E. Cassel, who had just established the National Bank of Egypt. He expressed a willingness to participate in the creation of the Anglo-Russian Bank, but using almost the same language as Lane, spurned Rothstein's proposal and demanded "something in the shape of inducements before we can get people to put their money into our venture of this kind.... Without this, it would be a forlorn hope to interest anybody."[71] But the Russian side could offer neither on a scale that would whet English appetites.

The negotiations for the Anglo-Russian Bank also drew in French and German bankers, whose mediation was used to push the proposal forward. However, the differences between the English and Russian sides proved irreconcilable. It is highly indicative that these unsuccessful negotiations to found an Anglo-Russian Bank involved the very same representatives of "the City" (London's financial district) who would consequently participate in the creation of the British oil empire. They included Lane, Suart, Gladstone, and Cassel—one of the future bosses of the English bank in Turkey (subsequently an instrument for seizing the largest oil reserves in the Near East that were opened on the eve of World War I).

Collapse of the proposal for an Anglo-Russian Bank did not mean an end to negotiations to expand English participation in the Russian oil business. It so happened that the St. Petersburg International Commercial Bank obtained those tracts on Bibi-Eibat that had so whetted the appetites of British firms. Once Rothstein became convinced that the negotiations for the Anglo-Russian Bank had reached an impasse, he sent his representative

A. Koch to London with the charge of selling these land tracts to Lane on profitable terms. The English eagerly began to negotiate, but also insisted that attention be given to their proclaimed "principles" of expanding "British participation in the Baku oil business." Lane further demanded that the Ministry of Finance guarantee prompt formal confirmation of the deal by the tsarist government.[72] Informed of all this, Rothstein immediately dispatched a representative to the ministry to press the case, and in fact, did obtain a positive answer.

Everything worked out so well that Lane decided to try to extort some additional advantages. That ploy infuriated Rothstein, who began to demand that the contract be signed immediately. Koch's position proved to be very complicated: during the day he had "to wage battle with these English scoundrels" and in the evening to read "rude telegrams" from St. Petersburg.[73]

Seeing that the English continued to stall, Rothstein gave free vent to his fury in a letter filled with abuse and invective. Under these conditions, he wrote, "there is no point in dealing with the gentlemen from London" and the St. Petersburg International Commercial Bank "has the right ... to retract its offers." "The entire group," he wrote, "are splendid people, but also a gang of extortionists, who want to stall as long as possible in order to have all the odds on their side and to exploit us." Rothstein was losing patience and ordered that an ultimatum be given to Lane: "Put it to these people directly: either/or. If they wish to make a deal, fine. If they want to act like swine, that's fine too. Send them to the devil and come home."[74]

The threats worked: the English signed the agreement to purchase the Bibi-Eibat oil fields. As a result, the St. Petersburg International Commercial Bank earned an enormous sum (approximately 5 million rubles), and the Shell group added a new and highly promising oil reserve to its Russian holdings. Altogether, in the late 1890s and early 1900s, the English succeeded in acquiring twenty-four oil-producing companies. In addition to the Baku region, where English capital owned eleven firms with a capital of 40 million rubles, the British also established a base in the other main oil-bearing region around Groznyi. There they established seven English firms with a capital of 11 million rubles; as a result, the Rothschilds—who had previously ranked first among foreign capitalists in Groznyi—found that they had dropped to second place. In these years too, the British formed the first companies in a number of other regions (for example, in Maikop) that would acquire growing significance in the future.

The entry of British capital into the Russian oil business was accompanied by grandiose plans. Compared to these designs, what had been achieved thus far was not very great, yet it was by no means insignificant. With good reason many thought that the English had achieved a position that approximated that of the two largest firms—Nobel and Rothschild. In any event, that was precisely the goal of the English entrepreneurs, who, as already noted, also made several attempts to reach an agreement with Mantashoff regarding the purchase of his enterprises. The role of intermediary in these negotiation belonged to K. Gulbenkian, subsequently one of the greatest brokers in the oil business. It was by accident that he first made Mantashoff's acquaintance in 1896—when he had journeyed by steamship from Constantinople to Egypt in search of refuge from the carnage in Turkey. For a short time, Gulbenkian even worked as Mantashoff's employee. Here he received his professional training before embarking on his dizzying career in business. Gulbenkian was soon an independent businessman in London and at the Parisian stock exchange. His brokerage abilities became legendary. Nevertheless, in the 1890s he did not succeed in persuading Mantashoff to conclude an agreement with the English entrepreneurs. This deal was simply beyond his capacity, although in the future he did draw the Baku millionaire into contractual relations with the English.

At the special conference convoked by the Russian government in early 1899, it was pointed out that the English were striving "to receive the most valuable oil-bearing lands on the Apsheron Peninsula, in which the entire Russian petroleum industry has an interest."[75] The tsarist government did not wish to make concessions on this scale, especially since its hopes for English loans had gone awry. In addition, that kind of decision was impossible because of the condition in Anglo-Russian political relations.

The German historian Hallgarten holds that the financial and economic ties between England and Russia from the late 1890s was a precondition for an improvement in the relations between the two countries and served as a kind of "financial prelude for the later Anglo-Russian entente."[76] That is all true enough, but something else must be taken into account: at the end of the nineteenth century England and Russia were still a long way from concluding an entente cordiale. This fact alone posed serious difficulties for British designs on Russian oil.

THE FAILURE OF THE GERMAN CARTEL SCHEME

Alongside the French and English, German capital also displayed a keen interest in Russian oil—an aspiration that also bore a certain political undercurrent. Indeed, it was made with the direct participation of political spheres, which made formal requests to Russian government to grant concessions to German capital.

The mounting international tension in Europe in the 1890s, the growing antagonism of Germany toward France, England, and the United States in the struggle for hegemony on the world arena—all this drove German diplomacy to attempt to restore good relations with Russia. This attempt was supported by the Berlin stock exchange, which offered new loans to Russia. At the same time, German banks took into account the favorable dynamics of the industrial boom of the 1890s and presented investment proposals to various Russian enterprises, including those in the petroleum industry.

The development of commercial and economic relations between Germany and Russia were advanced by the German side as an important political objective during a visit by Kaiser Wilhelm II in St. Petersburg in July 1897. Immediately after arriving in the Russian capital, the Kaiser met with Witte and awarded him the highest German honor (the Black Eagle), after emphasizing that this honor is given only to members of royal families and that the conferral on Witte constituted "an absolute exception." In the course of the negotiations, the Kaiser advanced a plan to create a commercial-political coalition of European countries against the United States. However, Witte personally demanded that this proposal be rejected as contradictory to the traditionally friendly relations between Russia and the United States.[77]

As for the German application to participate in the Russian oil business, that had already emerged before Wilhelm's arrival in St. Petersburg. The German financial domain proposed to Witte that they conclude an oil agreement, which would be directed against Standard Oil. As a result of this, Russia would receive a favorable tariff on the import of its oil products to Germany. Simultaneously, the largest bank in Hamburg (Warburg) wrote to Rothstein offering its services and expressing the hope of obtaining "the invaluable support of so influential a person." After learning of the negotiations by the Parisian Rothschilds for Mazut, Warburg sent Rothstein a query whether it was not possible to create a German enterprise for the exploration, extraction and distillation of oil.[78]

The German market consumed the largest quantity of kerosene in Europe, the consumption of oil fuel by railroads was increasing every year, and military circles contemplated proposals to convert naval ships to engines using petroleum fuel. At the same time, Germany did not have its own oil reserves and depended entirely on imports from Russia and the United States.

In view of the tense commercial and political relations with the United States, Berlin hit upon the idea of creating a bloc of European states, and it offered Russia advantageous terms for the import of its petroleum products.

In the spring of 1897 the United States adopted a protectionist tariff, closing the door to German products on the American market. In response voices in Germany demanded reprisals against American imports, above all the products from Rockefeller's Standard Oil.

Nowhere in Europe did the petroleum war become so intense as in Germany. To conquer the German market, Standard Oil had used all possible means. By the mid-1890s it controlled more than three-quarters of the kerosene imports. Business circles and the press both demanded measures to rein in the Rockefeller trust, to make access to the German market more difficult for American kerosene, and to encourage instead the import of petroleum products from Russia. These demands found a favorable response in the Reichstag. "Mr. Rockefeller dictates prices to us," declared one of the deputies, "we are in the grasp of an oil monopoly." To rid itself of American dominance, it was necessary to terminate trade with Standard Oil and to shift completely to the use of Russian oil. As the first step it was proposed to establish a state monopoly on the distillation of petroleum so that complete control could gradually be established over the kerosene trade—that important "factor of power" in world politics.[79]

The anti-American outbursts in the press and Reichstag appeared to preclude any compromise or ambivalence. And in their negotiations with the Russian side, the German representatives acted quite straightforwardly. The proposed privileges for Russian petroleum product applied above all to crude oil. The Germans thus made it clear that they themselves would engage in the refinement process. That is why the Nobel firm, once it had become familiar with the German proposal, expressed its opposition. It declared that, having acquired the capacities for oil distillation, Germany would no longer need to import kerosene from abroad and indeed would become an "extremely dangerous rival" for Russian oil industrialists.[80] The government endorsed the Nobel opinion. An instruction was given to the finance ministry's agent in Berlin (V. I. Timiriazev) to "restrain the German government from all such measures in this direction," for the import of Russian kerosene to Germany was "one of the serious concerns of the Ministry of Finance."[81]

Nobel submitted a memorandum to the Ministry of Finance on the question of Russo-German negotiations, which then became the subject of a special conference. The tsar was then informed of the progress of the negotiations. After examining these materials, the tsar issued this "royal" resolution: "I am very interested in the question being raised here."[82]

The trade in Russian kerosene was conducted by a Nobel branch (Naftaporta—Deutsche Russische Nafta Import Gesellschaft). Therefore, the memorandum sent to the German government indicated that the tsarist government assigned the highest significance to creating favorable conditions for this company.

Nobel's interests virtually determined the position of the tsarist government. Work on preparing the terms of the agreement and the negotiations themselves were conducted amidst daily contact with this firm. Nobel not only attended all these conferences at the Ministry of Finance, but also participated in preparing instructions for the ministry's agent in Berlin, V. I. Timiriazev: before Witte signed these instructions, agreement on their contents was first obtained from Nobel. It is not without interest that Timiriazev simultaneously held the position of chairman of the board of the Nobel firm Naftaporta.

It would seem that everything was going smoothly, but there was one hitch to the whole agreement: the Reichstag and the

German press expressed the fear that Germany would fall into a trap if Nobel concluded an agreement with Standard Oil. Each passing day brought new evidence that Germany and the United States had become irreconcilable rivals in the struggle for world commercial and industrial hegemony. Hence, deputies in the Reichstag and the press spoke out in the sharpest possible terms against American economic penetration, which was seen as a direct peril to German national interests. Summing up this mood, the German economist Prager cited the statement of a like-minded Austrian that "the American danger" will come "not in the form of troops and military fleets, nor in the form of new religious and political teachings," but as "crates of goods" that will "slash as sharply as a sword" and as "price lists" whose destructive power "is no less than that of the newest artillery shells."[83]

With the assistance of price lists and merciless price wars, the American Standard Oil endeavored to gain control of the German market. In the face of this threat, interested circles demanded that a reliable defense be raised up against Rockefeller's assault. Was it possible to count on Nobel, whom they constantly suspected of seeking to form a united bloc with Rockefeller? Press commentaries expressed the fear that Nobel might prove to be a "Trojan horse," who would assist the Americans in conquering the German market. State Secretary von Bülow and other high-ranking representatives of the German Ministry of Foreign Affairs requested Timiriazev to give a guarantee that Nobel would not make a deal with Rockefeller.

On the other hand, the attacks on Nobel provided a convenient backdrop for attempts by German capital to seize control over the petroleum trade for itself. It was with this goal that Warburg and other German banks undertook to obtain the creation of German oil enterprises in Russia, and their negotiations with Rothstein ran parallel with those between the two governments.

The German banks' proposal envisioned the organization of a central agency for a new firm to refine Russian kerosene in Hamburg and to create branches all across Germany. The proposed enterprise would seek to obtain oil fields in Baku, to construct an oil refinery on the Volga as well as storage tanks and other facilities, and also to acquire its own tank cars and oil tankers. Part of the output, especially the residual fuel oil, was to be sold in Russia itself.

In many respects, this proposal was reminiscent of the organization of the Rothschilds' Mazut firm, the negotiations for which were successfully concluded with Rothstein's assistance. The German entrepreneurs also expected to have the assistance of the St. Petersburg International Commercial Bank, emphasizing in their letters to Rothstein that they have "great interest" in their project.[84]

The German bankers tried to lure the Russian side with the prospect of a significant expansion of the kerosene business in Germany. But they found a rather cool reception in St. Petersburg, and thereupon turned to Paris in an attempt to reach an agreement via the Rothschilds. F. Wachenheim (a representative of a bank in Frankfurt-am-Main) was dispatched for that purpose to Paris, where he was to act in the name of a consortium that included his own bank, the Warburg Bank in Hamburg, and the Bleichröder Bank in Berlin. Although at first, these negotiations went successfully, they soon broke down when it became clear that the German terms simply did not suit the Rothschilds.

The latters' conduct so infuriated Wachenheim that he began to threaten to make an agreement with Mantashoff. Playing upon the contradictions among the various Russian groups, the representative of the German consortium succeeded in obtaining a resumption of the negotiations. In all possible ways he emphasized that he was seeking to organize an enterprise "without Nobel" and, apparently, this pleased the Rothschilds.[85] In any event, the deterioration in their relations with Nobel served to bring a resumption of the dialogue with Wachenheim.

Paris zealously followed the course of the negotiations regarding privileges for petroleum products in Germany, for it knew what role would be accorded to Nobel in realizing the proposed agreement. That is why, when Witte convoked a conference at the end of 1898 to make a decision on the introduction of a privileged railway tariff on kerosene exported to Germany, the Rothschilds' representative came out against the proposed reduction. He declared such a privilege to be "excessive," since it "would give an advantage to one firm at the expense of all others that export Russian kerosene to countries other than Germany."[86]

The German banks sought to exploit these disagreements, but the negotiations with the Rothschilds nonetheless came to naught. In April 1899 Wachenheim was forced to leave Paris. After consultations in Berlin he once more tried to realize his proposal, applying to the Rothschilds for an agreement. In the second half of May 1899 a representative of the German consortium arrived in St. Petersburg, where he visited the St. Petersburg International Commercial Bank and presented a proposal "to form a kind of cartel," which constituted a significantly expanded version of the original German plan.[87] It now proposed to establish a pan-European oil conglomerate against America's Standard Oil. This enterprise, in the conception of its authors, sought to control the oil trade not only in Germany, but in other European countries by relying upon "the centers, which Mr. Rothschild possesses in England, Belgium, and Holland for the sale of their products."[88] Along with that were plans to acquire oil fields in Russia, to construct an oil refinery, and also to participate in various Russian firms, including the Rothschilds' Mazut firm.[89]

It is possible that the negotiations would have continued longer. But the Nobel firm knocked the ground out from under the German consortium when it invited the Rothschilds to join the board of Naftaporta and to acquire 2 million Reichsmarks of its stock (as part of an increase in the company's capital from 1.5 to 5 million marks).[90] "I was afraid," Aron wrote, "that Baron Alphonse, who in many cases had reason to complain about the behavior of Mr. E. Nobel, would respond negatively to these attempts to enter negotiations and order me to refrain from these or, at least, would give an evasive response to the proposal of our competitor. Fortunately, this was not the case."[91]

That did not happen, for the two competitors were in fact very different. Whereas the Germans demanded that the Rothschilds make concessions (by admitting German capital to participate in its enterprises), the French side did not have to concede anything in making an agreement with Nobel—even as it obtained access to Naftaporta. This variant proved acceptable in all respects. Indeed, Nobel and the Rothschilds displayed an increasingly strong tendency to make a still broader agreement. As early as February 1898, Nobel proposed to conclude a cartel agreement about joint actions on the domestic market for the sale of residual fuel oil. This was precisely what the Germans had been seeking. However, whereas the German attempt had been immediately rejected as unacceptable, the Rothschilds responded quite differently to an analogous proposal from Nobel. "Your letter about the visit of Emmanuel Nobel," Aron wrote Rothstein, "is very interesting. You used all the resources of your experience with people and affairs in Russia and with a skillfulness that, to be sure, does not surprise me. You have forced our competitors to stop and reflect and, perhaps, have posted the landmarks for a future agreement, which is desirable from every point of view."[92] Indeed, the Nobel proposal did subsequently become the subject of negotiations, which after the turn of the century culminated in a cartel agreement called Nobmazut, whereby the two firms coordinated their trade policies on the domestic market.[93] Already in the early 1900s the Batumi excise supervisor noted that "the firms 'Nobel Brothers' and the company 'Mazut' occupy a completely exceptional position here, which bears the character of a monopoly."[94] By 1901, according to official statistics, their share amounted to 57 percent of the kerosene delivered to the domestic market via the Caspian Sea, as well as 43.5 percent of the residual fuel oil, and 67.5 percent of the lubricating oil.[95] Thus, the basis was laid for a union of the two largest oil industrial firms, which combined their forces for the purpose of monopolizing the oil business in Russia.[96]

The Nobel-Rothschild agreement meant of course that there was simply no place left for the German project. As soon as Nobel applied to Rothstein to render assistance in negotiations with the Rothschilds, it was clear that a more or less large-scale agreement was feasible, including one to resolve the question of the kerosene trade in Germany. As Rothstein observed: "If this combination is realized, Mr. Wachenheim no longer holds any interest for us."[97]

At the same time, Rothstein preferred to avoid a conflict with German banks and proposed that the Rothschilds surrender a quarter of their stock in Naftaporta. The Rothschilds agreed. However, in fact the German banks obtained less than a tenth (worth just 150,000 marks)—a pitiable crumb that could hardly compensate the Germans for the failure of their plan for an oil cartel.

The failure of the German banks to establish themselves in the oil business had a decisive impact on the fate of Russian-German governmental agreement. When the question of a favorable tariff on Russian kerosene was being decided in Berlin, it calculated that this measure could promote the plan for a German cartel. After that endeavor collapsed, it was clear that all the benefits from the preferential tariff would go to Nobel and the Rothschilds. Therefore, it was not surprising that the German press sounded the alarm, redoubling the force of their attacks on Nobel, whom they accused of conspiring with Standard Oil, adding that it acted with the sanction of the Ministry of Finance. When Witte was informed of this, he denounced the assertions in the German newspapers as "a brazen lie."[98] In December, 1898 the newspaper *Frankfurter Zeitung* published an official denial. But this had no effect. Attacks on Nobel, and simultaneously on the Ministry of Finance, continued to appear in the German press. They were inspired by German business circles which had been disillusioned by the failure of attempts to gain entrance to the Russian oil industry. On the other hand, Standard Oil's representatives also had a hand in unleashing these attacks on Nobel. Rockefeller was not in the least pleased by the prospect that his chief rival in the German kerosene trade secure advantages, to the detriment of his own company's interests. Therefore he did everything he could to kill this agreement. Nobel's so-called American "allies" not only staged newspaper attacks, but also submitted documentary evidence to the authorities of agreements that had been concluded with Nobel. The State Department also came to Rockefeller's assistance. As Timiriazev reported, "the U.S. Government, as I have learned, is making every effort to interfere in this agreement."[99]

After the plan for an oil cartel had foundered, German authorities began to back away from their commitment. And in St. Petersburg it was understood too that there were no chances of the agreement going into effect. Therefore, in referring to the violation of the agreed-upon dates for ratification, Witte requested the Minister of Foreign Affairs, Murav'ev, "to inform the German government" that "the agreement is losing its validity."[100] German leaders were verbally still reassuring the Russian representatives that the government "continues to make every effort to secure a Reichstag majority in favor of the new decree on kerosene."[101] But Witte categorically insisted on his own way: "Because of Germany's failing to meet the deadlines," the agreement had lost its validity.[102] That was the final word on the Russo-German negotiations for an oil agreement.

Later, in his memoirs, Witte asserted that he had understood the inevitability of war with Germany ever since the mid-1890s.[103]

There is no doubt that he was one of the first representatives of the tsarist elites to see the direction in which Russian-German relations were developing. It is entirely probable that this influenced his position in the negotiations for an oil agreement and, especially, his attitude toward attempts by German banks to claim a place in the Russian oil business. Witte did not display any interest whatsoever in the German proposals, and neither he nor indirectly through Rothstein did he give any promises in that regard. As we have seen, things stood quite differently in the negotiations with the French and English.

The German proposal was born in the heat of the struggle for Russian oil, when foreign capital was literally storming Russia to obtain advantageous oil concessions. This assault ended in the creation of many foreign companies, which in this period significantly intensified their position in the Russian petroleum industry. So far as the collapse of the Russian-German agreement and the unsuccessful efforts of the German banks are concerned, these can only be seen as a defeat for Germany and the ruin of its plans to participate in dividing up the Russian petroleum resources. All this had a certain underlying cause and testified to the irreversible, growing contradictions in Russian-German relations.

The intrusion of foreign capital into Russia and the competition of various groups in the struggle for Russian oil—all this developed against a background of mounting international tension, in which the oil wars played an important role.

NOTES AND REFRENCES

1. *U.S. Mineral Resources* (1905), p. 735.
2. P. V. Ol', *Inostrannye kapitaly v narodnom khoziaistve dovoennoie Rossii* (Leningrade, 1925), pp. 12-13, 19; L. Eventov, *Inostrannyi kapital v neftianoi promyshlennosti Rossii, 1874-1917* (Moscow-Leningrad, 1925), p. 37.
3. This characterization is taken from the American newspapers, which in June 1990 commented on Rothstein's arrival in the United States for negotiations on a loan. See *TsGIA SSSR,* f. 626, op. 1, d. 1734, l. 4.
4. Decoded telegram from Montebello (1897) in Ministere des affaires etrangeres, Archives diplomatiques (Paris, *Correspondance politique,* Russie NS, 51, p. 26).
5. S. Iu. Witte, *Vospominaniia,* 3 vols. (Moscow, 1960), *2,* p. 235.
6. Bompar to Buzhua (March 29, 1906) in *Correspondance politique, 57,* pp. 49-60.

7. G. W. F. Hallgarten, *Imperialismus vor 1914; die soziologische Grundlagen der Außenpolitik vor dem ersten Weltkrieg,* 2 Bde. (München, 1961), 1, p. 344.

8. Kovalevskii to Rothstein (not after 1895). In *TsGIA SSSR,* f. 626, op. 1, d. 349, 1. 15.

9. "Zapiska Rotshil'dov Peterburgskomu Mezhdunarodnomu Banku" (not earlier than late 1896) in *TsGIA SSSR,* d. 286, ll. 82-83.

10. See Iu. B. Solov'ev, "Franko-russkii soiuz v ego finansovom aspekte (1895-1900 gg.)," *Frantsuzskii ezhegodnik. 1961* (Moscow, 1962), pp. 163-206; idem, "Peterburgskii Mezhdunardnyi bank i frantsuzskii finansovyi kapital v gody pervogo promyshlennogo pod"ema v Rossii," *Monopolii i inostrannyi kapital v Rossii* (Moscow-Leningrad, 1962), pp. 377-407.

11. Spiro to Skal'kovskii (September 21/October 3, 1897). In IRLI, f. K. A. Skal'kovskogo, d. 8660 - XIX b41, 1. 2.

12. See B. V. Anan'ich, *Rossiia i mezhdunarodnyi kapital 1897-1914* (Leningrad, 1970), p. 16.

13. Rothschild to Rothstein (March 31/April 12, 1897). In TsGIA SSSR, f. 626, op. 1, d. 224, 1. 6.

14. Baer to Skal'kovskii (December 28, 1895/January 9, 1896). In LOII, f. 202, d. 21, kart. 3, ll. 252-253.

15. Hallgarten, 2, p. 323.

16. Aron to Rothstein (no earlier than August and no later than December, of 1896) in *TsGIA SSSR,* f. 626, op. 1, d. 285, ll. 310-311.

17. Aron to Rothstein (December 1/13, 1897). In *MKNPR,* p. 214.

18. Aron to Rothstein (December 14, 1896). In TsGIA SSSR, f. 626, op. 1, d. 230, 1. 6.

19. Feigl' to Rothstein (March 12/24, 1896). In TsGIA SSSR, d. 352, ll. 6-7.

20. Rothschild to Rothstein (May 1, 1987). In *MNKPR,* 194-195.

21. Ibid.

22. Aron to Rothstein (April 21, 1897). In *MNKPR,* p. 193.

23. Aron to Rothstein (July 30, 1897). In *MNKPR,* pp. 205-206.

24. A. Aron to Rothstein (August 26, 1897). In *TsGIA SSSR,* f. 626, op. 1, d. 224, 1. 40.

25. A. A. Davidov to Rothstein (August 16/28, 1897). In *MKNPR,* p. 207.

26. Memorandum of G. A. Hanotaux (October 23, 1897). In *Documents diplomatiques français* Paris, ser. 1, 13, pp. 565-566.

27. Rothschild to Rothstein (April 12, 1897). In *TsGIA SSSR,* f. 626, op. 1, d. 224, ll. 5-6.

28. Aron to Rothstein (November 11, 1897). In *MKNPR,* p. 211.

29. Aron to Rothstein (November 12, 1897). In *TsGIA SSSR,* f. 626, op. 1, d. 286, 1. 60.

30. Rothschild to Rothstein (May 3, and November 18/30, 1897). In *TsGIA SSSR,* f. 626, op. 1, d. 224, ll. 14-15, 70.

31. Circulat letter and list of stockholders of Mazut (March 28/April 9, 1898). In *TsGIA SSSR,* f. 626, op. 1, d. 363, ll. 36, 2.

32. Reports from Chambers (dated March 7, 1898 and February 23, 1899). In *U.S. Consular Reports, 57,* p. 42 and *60,* p. 223.

33. Report from the consul Smith (dated April 29, 1898). In *U.S. Consular Reports, 57,* p. 472.

34. G. Vilenkin to V. N. Kokovtsov (April 9/21, 1898). In *U.S. Consular Reports, 57,* p. 472.

35. Aron to Rothstein (November 11, 1897). In *TsGIA SSSR,* f. 626, op. 1, d. 230, l. 166.

36. See, AVPR, f. Posol'stvo v Londone, op. 520, d. 873.

37. E. Hubbard to Witte (April 22, 1898). In *Istoricheskii arkhiv* (1960) *6,* p. 82.

38. Witte to Hubbard (13/25.V.1898) in ibid., pp. 85-86.

39. See Witte, 2:170.

40. B.V. Anan'ich, "Russkoe samoderzhavie i vneshnie zaimy v 1888-1902 gg." In *Iz istorii imperializma v Rossii* (Moscow-Leningrad, 1959), 189-192.

41. *Kaspii,* March 12, 1898; G. Vilenkin, *Finanosvyi i ekonomicheskii stroi sovremennoi Anglii* (St. Petersburg, 1902), pp. 8, 61-62.

42. M. Ia. Gefter, L. E. Shepelev and A. M. Solov'eva, "O proniknovenii angliiskogo kapitala v neftianuiu promyshlennost' Rossii (1898-1902 gg.)," *Istoricheskie zapiski* (1960) *6,* p. 77.

43. The letter and memorandum from Grand Duke Aleksandr Mikhailovich were included in a report for a Special Conference that was prepared by the Mining Department of the Ministry of Agriculture and State Domains (April 20/May 2, 1898). In *TsGIA SSSR,* f. 20, op. 7, d. 179., ll. 57, 60.

44. Ibid., l. 57.

45. Osobyi zhurnal Komiteta ministrov (May 19/31, 1898). In *TsGIA SSSR,* f. 1263, alfavit za 1898 g., d. 5331, ll. 213-214.

46. Ibid., l. 215.

47. Ibid.

48. Ibid.

49. Ibid., l. 213.

50. Ibid., ll. 215-217.

51. See Zhurnal Osobogo soveshchaniia (June 3/15, 1898). In *MKNPR,* pp. 219-228.

52. Anan'ich, pp. 186-87, 192; Memorandum of Jusseran (February 20, 1897). In *Documents diplomatiques français (1871-1914),* I série, *13* (Paris, 1953), pp. 215-216.

53. For details see: Iu. B. Solov'ev, "Protivorechiia v praviashchem lagere Rossii po voprosu ob inostrannykh kapitalakh v gody pervogo promyshlennogo pod"ema," *Iz istorii imperializma v Rossii* (Moscow-Leningrad, 1959), pp. 371-388.

54. Ibid., p. 375; *Moskovskie vedomosti,* January 24, 1899.

55. Solov'ev, "Protivorechiia," pp. 382-383; Gefter and Shepelev, p. 78.

56. Solov'ev, "Protivorechiia," pp. 373, 377-378, B. F. Brandt's study was entitled *Inostrannye kapitaly. Ikh vliianiia na ekonomicheskoe razvitie strany* Vol. 1 (St. Petersburg, 1898).

57. Evidence for this is to be found in the notes and diary of the state secretary, A. A. Polovtsev, *Krasnyi arkhiv* 46 (1931), p. .119.

58. Witte's memorandum was first published in an English translation by the American historian, Theodore von Laue, in *Journal of Modern History,* 36 (1954, pp. 67-74, based on a copy in the Library of Congress. The Russian text, found in the archive of Nicholas II in TsGIAgM, was published with related documents by I. F. Gindin in "Ob osnovakh ekonomicheskoi politiki tsarskogo pravitel'stva v kontse XIX—nachale XX v." in *Materialy po istorii SSSR,* 6 (Moscow, 1959), pp. 159-222. The author has found still another copy of the memorandum in the archive of the Ministry of Foreign Affairs (AVPR, f. II department, I-I, r.2, op. 358, d. 36, 11. 1-12).

59. See the memorandum of March 14/26, 1899. In AVPR, f. II department, 1-I, r.2., op. 358, d. 36, 1. 14; see also the minutes of Murav'ev speech at the session of the Council Ministers on March 17/29, 1899 in *Materialy po istorii SSSR, 6,* p. 204, 220.

60. See the zhurnal Osobogo soveshchaniia (no later than May 1/13, 1899). In *TsGIA SSSR,* f. 40, op. 2, d. 114, l. 83.

61. Ibid., l. 92.

62. Solov'ev, "Franko-russkii soiuzz," p. 196.

63. Ibid.' Spitzer to Rothstein (April 15/27, 1899). In *TsGIA SSSR,* f. 626, op. 1, d. 87, 1. 121.

64. R. Salisburgy to N. O'Connor (January 27, 1898). In *British Documents on the Origins of the War, 1898-1914.* Edited by G. R. Gooch and H. Temperley (London 1927), *1,* p. 5.

65. O'Connor to Salisbury (July 12/24, 1898). In *Foreign Office,* 65/1555 (Public Record Office, London).

66. Ibid.

67. B. A. Romanov, *Ocherki diplomaticheskoi istorii russko-iaponskoi voiny. 1895-1907* (Moscow-Leningrad, 1956), pp. 96-103.

68. Minutes of the statement by A. S. Ermolov at the Special Conference (no later than the end of February 1899). In *TsGIA SSSR,* f. 37, op. 66, d. 2658, ll. 40-41.

69. Ibid., ll. 43-44.

70. Solov'ev, "Peterburgskii mezhdunarodnyi bank," p. 378.

71. E. Cassel to E. Noetzlin (May 13, 189). In *TsGIA SSSR,* f. 626, op. 1, d. 247, l. 43.

72. See, Rothstein to Koch (September 25/October 8, 1900). In *TsGIA SSSR,* d. 371, ll. 85-86.

73. Rothstein to Koch (September 24/October 6, 1900). In ibid., l. 84.

74. Rothstein to Koch (September 26/October 9, 1900; September 27/ November 11, 1900; October 10/23, 1900). In *TsGIA SSSR,* 11. 89, 95-96, 187.

75. Journal of the Special Conference (late February and early March, 1899). In *TsGIA SSSR,* f. 37, op. 66, d. 1658, l. 41.

76. Hallgarten, 1:548.

77. Witte to M.N. Murav'ev (November 12/24, 1897). In *TsGIA SSSR,* f. 20, op. 992, ll. 105-106.

78. M. Warburg to Rothstein (January 17, 1897). In *TsGIA SSSR,* f. 626, op. 1, d. 136, ll. 77-78.

79. A. S. Erusalimskii, *Vneshniaia politika i diplomatiia germanskogo imperializma v kontse XIX veka* (Moscow, 1948), p. 349.

80. See, Kovalevskii to V.I. Timiriazev (December 5/17, 1897 g.). In *TsGIA SSSR,* f. 20, op. 7, d. 195, ll. 88-89.

81. Ibid., ll. 89, 93.

82. See, A.P. Izvol'skii to Murav'ev (May 23/June 4, 1898) in *AVPR,* f. II department, 1-5, op. 407, d. 739, 1. 1.

83. M. Prager, *Die amerikanische Gefahr* (Berlin, 1902), pp. 17-18.

84. Warburg to Rothstein (August 21/September 12, 1898). In *TsGIA SSSR,* f. 626, op. 1, d. 136, ll. 170-171.

85. See, Timiriazev to Kovalevskii (August 15/27, 1898). In *TsGIA SSSR,* f. 20, op. 7, d. 195, l. 252.

86. The minutes of the statement by the representative of the Caspian and Black Sea Co. at the conference of December 16/28, 1898. In *Materialy po istorii SSSR,* 6, p. 137.

87. Rothstein to Aron (May 30/June 11, 1899). In *TsGIA SSSR,* f. 626, op. 1, d. 370, ll. 18-20.

88. Ibid.

89. Aron to Rothstein (June 15, 1899). In *MKNPR,* p. 235.

90. J. Wachenheim to International Bank (April 22/May 4, 1898). In *TsGIA SSSR,* f. 626, op. 1, d. 270, 1. 15.

91. Aron to Rothstein (June 30, 1899). In *MKNPR,* p. 237.

92. Aron to Rothstein (March 8, 1898). In *MKNPR,* p. 217.

93. *MKNPR,* pp. 700-702; P. V. Volobuev, "Iz istorii monopolizatsii neftianoi promyshlennosti dorevoliutsionnoi Rossii (1902-1914 gg.)," *Istoricheskie zapiski,* 52 (1955), pp. 89-94.

94. Report for 1899 from K. Simchenko, the excise supervisor of the eighth district of the Office for Excise Collections in the Trans-Caucasus District and the Trans-Caucasus region. In *TsGIA SSSR,* f. 270, d. 822, l. 20.

95. *Obzor bakinskoi neftianoi promyshlennosti za 1901* g. (Baku, 1902), chast' *2,* pp. 237-239.

96. Volobuev, pp. 82-83; A. A. Fursenko, *Neftianye tresty i mirovaia politika 1880—2* gg.—*1918* g. (Moscow-Leningrad, 1965).

97. Rothstein to Aron (June 15/27, 1899). In *TsGIA SSSR,* f. 626, op. 1, d. 370, l. 32.

98. See, "Rezoliutsiia Vitte na perevode stat'i *Frankfurter Zeitung.*" In *Materialy po istorii SSSR, 6,* pp. 155-156.

99. Timiriazev to Kovalevskii (January 9/21, 1899). In *Materialy po istorii SSSR,* p. 143.

100. Witte to Murav'ev (January 7/19, 1900). In *Materialy po istorii, SSSR,* pp. 153-153.

101. N. D. Osten-Saken to Murav'ev (January 10.22, 1900). In *AVPR,* f. II department, 1-5, op. 407, d. 740, l. 44.

102. Witte to Murav'ev (January 14/26; January 25/February 6, 1900). In *AVPR,* ll. 46, 55.

103. Witte, *Vospominaniia, 1,* 546-547; B. V. Anan'ich and R. Sh. Ganelin, "Opyt kritiki memuarov S. Iu. Vitte (v sviazi s ego publitsisticheskoi deiatel'nost'iu v 1907-1915 gg.)," *Voprosy istoriografii i istochnikovedeniia istorii SSSR* (Moscow-Leningrad, 1968), p. 373.

Chapter III

The "Kerosene Wars" in the Far East and the Formation of Royal Dutch Shell

The chief arenas of the first oil wars were those places where the interests of Standard Oil and Russian oil magnates came into conflict. In the 1890s a "second front" opened after the discovery of oil in the Dutch (Netherlands) Indies and rapid development of its production, which soon began to pour into the far eastern markets, above all into China and Japan. The sale of kerosene in the Far East was a profitable business and had excellent future prospects. Hence, the appearance of a new competitor greatly complicated the situation there. In 1890, on the island of Sumatra, the Royal Dutch company was formed, which in the future was fated to become one of the largest oil monopolies of the world.

Notwithstanding the seriousness of the Russian-American oil conflict, Rockefeller and the Russian oilmen had worked out a form of coexistence that ensured a certain equilibrium in their mutual relationship. That order, however, now came under a severe test. A virtually unknown oil company, in a short time, emerged as a major threat to the powerful trusts. By the end of the 1890s several firms were operating on Sumatra and other islands of the Dutch Indies, but by far the most successful was the Royal Dutch company.

The import of kerosene from the Dutch Indies to China was begun in 1893. Three years later it was exported to all the ports of China and had attained the impressive volume of five million gallons. Within a year that amount had doubled; a year later it had increased another than five-fold.[1]

By this time, 1898, the volume of kerosene exported from the Dutch Indies to China equalled that of Russian producers and over half that of the Americans.[2] It proved very difficult to compete with the Dutch entrepreneurs. The richness of the oil reserves, the geographic proximity of the Dutch Indies (located alongside China), the cheap labor force—all this gave them a great advantage and made it possible to sell at cheap prices.

THE RISE OF THE ROYAL DUTCH COMPANY

Having established its place in the Far Eastern market, the Royal Dutch Co. began to receive high profits. Its dividends rose accordingly. In 1894 they amounted to 8 percent, but in subsequent years they increased severalfold, rising to 44 percent in 1895, 46.5 percent in 1896, and 52 percent in 1897. The price of its stock rose sharply. Therefore, when the company's capital was increased in 1897 from 1.7 million to 5 million florins, the earnings from the sale of stock (reaching 13 million florins) was several times their nominal value.[3]

The success of the Dutch Indies' oil industry had a particularly strong impact on Standard Oil. This was the strongest blow to its interests since the emergence of Russian oil companies in the 1880s. As was earlier the case in Russia, Standard Oil for a long time held a monopoly on the delivery of kerosene to the islands of Sumatra, Java, and Borneo. With the growth of local production, the import of foreign kerosene came to an end, and then it began to threaten the sale of American kerosene on the world markets. "The export of oil from the Netherlands Indies," confirmed an expert on petroleum in the U.S. Government, "is quite significant, and its sphere of influence encompasses all of Asia and Australia."

How did Rockefeller react to the appearance of so dangerous a rival? What measures did Standard Oil take to defend its position on the far eastern markets? This question became the subject of a special discussion at Rockefeller headquarters, where a decision was taken to involve the Dutch oil men in the American plan to carve up the world into spheres of influence. The examination of this question coincided with Russian-American negotiations on dividing the world market. According to Rockefeller's scheme, the Royal Dutch company should be included in the proposed agreement on the condition that it first conclude a bilateral contract with the American company.

Having completed negotiations in mid-1895 with Nobel and the Rothschilds, the head of the foreign section of Standard Oil, William Libby, invited the head of the Royal Dutch company, August Kessler, to meet with him in Paris. In the course of this meeting, he proposed to conclude a contract so that the Dutch company become "interested in our business." In that event, explained Libby, "the Royal Dutch Company would be assured of the support of this powerful concern, whose resources, financial as well as technological, were very great."[4]

The representative of the "powerful concern" endeavored to persuade Kessler that Rockefeller does not make it his goal to subordinate the Dutch Company. Kessler, nevertheless, suspected that Standard Oil simply wanted to swallow up his company. He therefore did not respond to the American proposition and left Paris.[5]

Rockefeller did not give up. A commission (including W. Libby, K. Lufkin, and D. Fertig) was formed with the task of continuing the negotiations and commenced work in early 1897. Lufkin and Fertig were dispatched to the Dutch Indies, while Libby set off for Europe, where he intended to become acquainted with the financial reports of the Dutch company, and also to meet with its representatives, insofar as he has already conducted negotiations with them.[6]

During a conversation with Kessler in Sumatra, Lufkin expressed his regrets, saying "Would it not be a pity that two such big concerns as you [sic] and our own should not go together." The representatives of Standard Oil put pressure on Kessler and tried to persuade him to form an alliance. At the same time, they began negotiations with the other Dutch company, Moeara Enim, "the most significant" company after Kessler's firm. On the one hand, the Americans hoped to make Kessler more tractable, for they had declared unequivocally that they would like to see in Moeara Enim a worthy rival for the Royal Dutch company. On the other hand, they began to investigate the possibility of acquiring Moeara Enim, having come to the conclusion that it was essential to absorb one of the successful Dutch firms.[7]

Once these exploratory steps had been taken, the heads of Standard Oil convoked a meeting in London to consider a further plan of action. "Every day makes the situation more serious and dangerous to handle," declared a memorandum on the conference. "If we don't get control over the situation soon, the Russians, Rothschilds or other party may."[8]

So they turned again to the Royal Dutch company. The American side proposed that the Dutch firm increase its stock fourfold, but on condition that all the new stock be purchased by Standard Oil. However, the American proposal was rejected once again. Then Rockefeller turned to the English company, Shell, which by this time owned oil fields on Borneo. But when this also failed to bring results, the Americans next decided to renew negotiations with Moeara Enim.

This company owned rich oil fields, but suffered from a constant shortage of financial resources. The American representatives proposed to give Moeara Enim the necessary wherewithal, and on February 12, 1898 signed a contract with the firm: Rockefeller was to give the company 6 million florins in exchange for 60 percent of its stock and two-thirds of the votes on the governing board.

News of this deal produced a sensation in the Netherlands. The price of Moeara Enim's stock on the American exchange leaped from 130 to 185 points, while that of the Royal Dutch company plummeted from 605 to 480 points.[9]

The American consul service regarded the agreement with Moeara Enim as "a most important step" by Standard Oil in its efforts "to enter the oil producing field in this part of the world."[10] The Russian consul, N. Shuiskii, made a similar assessment and expressed the fear that "the Americans will not stop with this, but will go further."[11] In the event that the deal with Moeara Enim obtained the necessary approval from government authorities in the Netherlands, competition with the American oil trust in the Far East would become practically impossible.

That government approval is precisely what Royal Dutch now sought to forestall, as its representatives sounded the alarm about the dangerous ramification of the Standard Oil-Moeara Enim agreement. Stirring up dissatisfaction in business circles and inciting a campaign in the press, they succeeded in winning the Dutch government to support their position. As a result, when the management of Moeara Enim filed for government approval of its agreement with Standard Oil, it was told that "American capital is not desired in the Netherlands Indies."[12] "The interests of the Netherlands and of the Indies," said the colonial minister, "would have been incomparably better served if the principal petroleum companies in the Indies had endeavoured to cooperate." He added that "when contact had to be sought, for purposes of sale abroad, with powerful foreign combines such as the American and the Russian Petroleum trusts, the Netherlands Indies industry would then have been able to negotiate upon equal footing and it would not have been necessary for it to become the obedient handmaiden of one of these trusts."[13]

Hence the agreement with Moeara Enim did not receive the approval of the Dutch government, and the Royal Dutch company celebrated its victory. The Hague's decision was also greeted with satisfaction by interested parties in Russia, for, as the Russian consul in Tientsin (Shuiskii) observed, the export of kerosene from Russia would have awaited a "sad future" in the event of Rockefeller's success.[14] A similar attitude prevailed among the English, for whom the oil trade acquired increasing significance with each passing year. The British consul in Amsterdam, Robinson, with unconcealed glee noted that the Dutch government "categorically" refused to grant concessions to companies "that are under the control of the monstrous American monopoly."[15]

For understandable reasons, the American reaction was quite different. After Rockefeller appealed to the American government to protest the decision of Dutch authorities, the State Department authorized its consul in the Dutch Indies to see that no kind of discrimination against American trade and American interests be tolerated. Simultaneously, it directed its envoy in The Hague to make a vigorous protest to the Dutch government.[16]

All this was done. But it did not lead to any concessions on the part of the Dutch, and the United States Government was powerless to do anything on Rockefeller's behalf. Both because of domestic considerations (opposition to trusts was mounting inside the United States itself) and because of the international situation (the Spanish-American War had just commenced), it was not in the United States' interest to provoke a crisis in its relations with the Netherlands. Rockefeller's agreement with Moeara Enim was therefore buried and forgotten.

To compensate for the weakness of its political support, the Rockefeller oil trust declared a merciless price war on the Dutch companies. After opening an agency to sell American kerosene on Java, Standard Oil began to deliver massive quantities of oil from the United States. To destroy its rivals, the company sold at dumping prices and, according to the American consul S. Everett, "is doing business at a loss."[17] Rockefeller's representatives began to buy storage facilities in port cities in pursuit of the same goal—to cause difficulties for the Dutch companies. Analyzing the tense situation in the Dutch Indies, Standard Oil endeavored as far as possible to prevent the further development of local industry.[18]

The Dutch companies suffered heavy losses. The price war led to a sharp fall in its stock, leaving some firms teetering on the verge of bankruptcy. The dividends of the Royal Dutch Co. began to tumble downward—from 52 percent in 1897 to 6 percent in 1898 and 1899. The company simply was not generating enough income and seemed on the brink of ruin and bankruptcy. "The Americans declare aloud," wrote the Russian consul in Batavia, "that the future is in their hands and that, with the capital at their disposal, at any time they can buy anything or anyone they please."[19]

Once again, however, Rockefeller miscalculated: the Dutch government permitted the companies to issue privileged stock. It added the proviso that such stock could not be sold to foreigners, the purpose being, of course, to prevent the Americans from buying up any of the new stock issues. It therefore enabled the Royal Dutch Co. to obtain the necessary finances and, indeed, the subscription significantly exceeded the announced sums. Moreover, the Dutch firm was also aided by the Parisian Rothschilds, who offered a monetary loan to use in the struggle with the American trust.

Although the Royal Dutch Co. had beaten back Rockefeller's attack, its leaders understood that it simply lacked the resources, over the long term, to wage war with the Americans. The Rothschilds' loan opened the possibility of obtaining money in France. But the unfolding of events suggested still another route—unification with British capital, which played so important a role in the petroleum trade in the Far East. Moving in this direction, the Dutch firms became virtually unassailable for Standard Oil. Rockefeller encountered here, for the first time, the future leader of the Royal Dutch Co., Henry Deterding, subsequently one of the greatest oil magnates of the world. The most bitter fights still lay in the future, but precisely at the turn of the century a conflict emerged that would later grow into a worldwide battle of the oil trusts.

By throwing down the gauntlet to the Dutch companies, Rockefeller inadvertently drove them into a union with London City. At the end of the 1890s the first steps were taken for a rapprochement with British capital that would eventually culminate the formation of a new oil concern that would become a most important factor in oil politics in the international arena.

The height of the battle for oil in the Dutch Indies coincided with a more intensive search by British companies for oil reserves. After The Hague vetoed the agreement between the Americans and Moeara Enim, the British company Shell offered its financial support to the Dutch firm. Although negotiations at first fared badly, the two firms finally signed an agreement in January 1899 that tripled the company's capital and doubled its production. The English acquired the controlling share of stock, but the two parties took measures to avoid "provoking the suspicion of the Dutch government" and thus the fate of the Rockefeller agreement.[20] The agreement predictably provoked great indignation from the American side. Did this new contract not prove that Standard Oil was being treated differently from Shell? The American diplomatic service saw the Anglo-Dutch agreement as "a clear case of discrimination against American interests."[21]

Representatives of the United States deemed it to be beyond any doubt that the agreement aimed "to keep American oil out of the market" of the Dutch Indies and had been framed "evidently in opposition to the Standard Oil Company." Not without reason they suspected that "a strong combination" was "working hand in glove" with the Parisian Rothschilds, and that "it looks like a combination of the Russian and Sumatra petroleum interests against American oil." The American representatives found proof for their views in the behavior of the Royal Dutch Co., which had also begun negotiations with English firms.[22]

Deterding first hit upon the idea of uniting with British capital in 1896, when, after becoming A. Kessler's right-hand man, he made an inspection trip of ports in the Far East and saw how England's powerful political influence helped to support her commercial interests. In early 1897 Deterding visited London and then went to India, where he signed an agreement with the English company of Gladstone (already known to us from its participation in Russian oil enterprises) to collaborate in the Far East. This was the first, modest contact, but the Dutch company planned to increase it in the future.

The question of an agreement with the English became particularly urgent in 1899, when, as a result of Rockefeller's price war, the Royal Dutch Co. experienced a serious deterioration in its position and its oil production fell off sharply. The company's difficulties were aggravated by the agreement between Moeara Enim and Shell. Therefore, as soon as the Gladstone group proposed to create a syndicate for combined sale of kerosene in the region east of the Suez Canal, the offer immediately became an object of intensive business negotiations.

The Royal Dutch Co. could expect to receive whatever quantity of kerosene it needed from the English oil firms in Baku. Without waiting for the end of negotiations, Kessler notified his agents in India and China that "we shall start importing Russian oil in bulk to the markets of the East."[23]

The Dutch company believed that it could survive the struggle with the Americans if it joined forces with the English firms. For his part, Gladstone pursued one further goal. By means of the union with the Royal Dutch Co., he hoped to shatter the power of his competitors in the Far East—the English company of Shell and the Parisian Rothschilds. He proposed to create a combined Anglo-Dutch fleet of oil tankers, which would operate on a coordinated schedule, and to organize a system of agents in countries of the East that would counterbalance the widespread network of Shell.

The scheme was grandiose, but the two sides agreed on terms in May 1899 to form a trade syndicate for joint operations in the Far East. To judge from the volume of resources invested (40,000 £ Sterling), the deal was quite modest and bore only a faint resemblance to Gladstone's original proposal. However, its anti-Shell thrust was unmistakable. That is why Samuel, as soon as he learned about the creation of this syndicate, ordered an increase in the purchases of Russian kerosene—in hopes of undermining his adversaries' plans. Nevertheless, he preferred to avoid a price war with the Dutch company and gave it to understand that he was prepared to negotiate. "The East is quite big enough," he declared, "for both the Royal Dutch and the Shell line."[24]

That was all said with the intention of it reaching Kessler's ears. But on December 14, 1900 Kessler died, bequeathing the post of company chairman to Deterding, who now became the sole master of the Royal Dutch Company.

That change of leaders marked a shift toward rapprochement with the Shell Co. In early 1900, Gladstone went to the Netherlands in order to extend his personal congratulations to Deterding, but received a chilly reception. The term of the agreement with the Royal Dutch Co. had expired, but the prospects of closer ties between the Dutch firm and Shell became more and more real. That is precisely what Deterding was seeking. F. Lane, who represented simultaneously the interests of both Shell and the Parisian Rothschilds, worked toward the same end. Deterding later wrote in his autobiography that "Frederick Lane was the cleverest man I have known in all my experience."[25]

They found a common tongue, though by no means immediately. Several years before the beginning of these negotiations, Lane had come to the conclusion that "the rising competition of kerosene from Sumatra and Java" was as dangerous as "the competition of American kerosene."[26] However, he changed his opinion once he decided that the sole solution was an agreement with the Dutch firms. Therefore, when Deterding appealed to him in the fall of 1901 to supply "good offices" in reaching an agreement with Shell, the Englishman was ready to undertake this mission. Indeed, Lane went even further, developing a proposal for a business concern that would unite all those—*except* Standard Oil—with an interest in the kerosene business in the East. The key to the concern's formation was first to unify the oil companies in the Dutch Indies and Russia. Deterding began to work on the Dutch Indies side, while Lane promised to handle the Russian side with the aid of the Rothschilds.

Although there were many obstacles to the Anglo-Dutch agreement, their mutual interests took the upperhand. On May 17, 1902 Deterding signed an agreement with Samuel to create a commercial concern called Shell Transport and Royal Dutch Petroleum Co.

The campaign to join forces with the Russian firms did not fare so well. The negotiations in Baku and St. Petersburg, which aimed at creating a new export syndicate, ended in failure. When the

Rothschilds expressed a readiness to join the Anglo-Dutch concern as a third party, Samuel took exception on the grounds that this was simply insufficient. Deterding, however, took a somewhat different view. For his part, he saw no chance of realizing the proposal without the participation of the Russian firms. He also expected to become executive director of the future concern and calculated that it would be much easier to manage and maneuver if the new conglomerate included not two, but three participants. On those grounds Deterding made the most energetic attempt to persuade Samuel of the desirability of including the Rothschilds as a third party. "I am very surprised," he wrote Lane, "that we do not go on with Paris independent of the smaller producers and, if necessary, independent of Mantashoff." He warned that "we must act quickly, if we want to combine with the Russians at all." "Once we combined with Rothschild," he declared, "everybody knows that we hold the future, but we cannot do without their name."[27]

He forced Samuel to agree and, on June 27, 1902, the Rothschilds (along with some other participants) provisionally agreed to create the concern "Asiatic Petroleum Co." and, a year later, they signed the final contract. Asiatic's capital amounted to 2 million £ Sterling and was divided into three equal parts among Deterding, Samuel, and the Rothschilds.

They made further efforts to attract Russian firms, but did not succeed in attracting Nobel to their association. The stumbling block was Nobel's insistence that he alone be given a quota for the kerosene that was significantly larger than what had been allotted for all the participants on the Russian side. A further obstacle was Nobel's contract with the American Colonial Oil Co., a branch of the Rockefeller trust, which laid claim to the very same markets in which Asiatic planned to operate.

The negotiations with Mantashoff ran more smoothly and, with Gulbenkian's mediation, led to a formal agreement. However, it did not prove long-lasting, chiefly because Mantashoff had joined Asiatic reluctantly.[28]

Despite all these problems, Deterding continued to seek agreement with the Russian firms. Acting through the Rothschilds, two years later he succeeded in obtaining Nobel's inclusion when the Parisian bankers agreed to surrender part of their delivery quota to Nobel.

The formation of Asiatic, which united English, Dutch, French, and part of the Russian oil interests, also reflected political realities, embodied in the 1904 Anglo-French Entente and later the 1907 Anglo-Russian agreement. As for the difficulties of obtaining the involvement of the Russian firms, these too can be ascribed, in part, to the international situation and the situation that emerged for Russia in the Far East just before and then during the Russo-Japanese War.

At the same time, formation of the Asiatic Co. marked an important step towards the creation of a world-wide oil monopoly. Participants in the concern had to conduct their Far Eastern trade through its mediation and in accordance with the mutually established terms. Geographically, the agreement encompassed literally half the globe—not only Asia, but also Australia and Africa. By the time that Asiatic was formed, the trade in liquid fuel had acquired great significance. Therefore, along with the sale of kerosene, the concern planned to establish special oil depots for oceanic ships throughout various parts of the world. "The new company in Asia," observed the contemporary press, "will doubtlessly take over the delivery of liquid fuel for oceanic vessels along the entire line from Australia to northern Europe. It already controls a significant part of this trade, which is being created at the expense of coal."[29]

The East became Asiatic's sphere of operations, but the most important result of this agreement consisted in the consolidation of European oil interests in the face of an intensification of American expansion. It is highly indicative that Deterding, once he had attained (with the Rothschilds' help) life-tenure as the head of the concern, decided to transfer his office to London. Formally, the management of the Royal Dutch Co. was still in The Hague, but in fact its fate was decided in London—and indeed, from this time its very name was anglicized (as Royal Dutch Co.) At the time, the capital of the British empire was the center for controlling trade and finances not only in the East, but in other parts of the globe. From the end of the 1890s London encountered a powerful rival—New York. Wall Street strove to replace London as the financial capital of the world. Still, although London had suffered some large losses in its economic competition with America, it nevertheless remained the world's commercial and financial center, as it was described at the time. From this base Deterding, who hereafter closely associated his fate with the imperial interests of England, prepared for new battles over the world's oil riches.

THE ROCKEFELLER-NOBEL BLOC: THE AGREEMENT FOR A RUSSIAN-AMERICAN BANK IN THE FAR EAST

As the contemporary press weighed the impact of the formation of Asiatic, it predicted that the next step would be an agreement with Rockefeller. Indeed, as soon as an agreement had been reached to create a concern for the kerosene trade in the Far East, Deterding telegraphed New York that he was ready to sign an agreement for joint operations. He then set off for America and began negotiations. But they did end in success.

Standard Oil took the opposite course: Rockefeller had sufficient power to follow an independent course and not become a draft-horse for Asiatic. In addition, he had his own plans to win over the Russian oil industrialists. True, Standard Oil put its hopes on Nobel, not the Rothschilds; it had already established ties with the Nobel firm and planned to strengthen these further.

Rockefeller had always attached special significance to a rapprochement with Nobel. After the failure to create a union of Baku kerosene manufacturers, this question became the subject of a special discussion at Standard Oil's headquarters at 26 Broadway. Rockefeller's and Nobel's representatives met regularly, as a rule choosing some neutral point in Europe for their meetings. If absolutely necessary, Rockefeller and Nobel could also communicate through Standard Oil's agent in Batumi (the American consul Chambers) or through the branch offices of both countries in various European countries.

Both companies adhered to common policy. People even spoke of a Rockefeller-Nobel bloc, whose existence no one ever questioned. In his old age Rockefeller, in a conversation with his biographer Inglis, admitted as much himself. He explained his relationship to the magnates of Russian oil as an attempt to eliminate competition. "The Russians had a sufficiently broad, reasonable approach to understand this," he declared. Confidential papers of Standard Oil later revealed that, in the view of Rockefeller's people, the most important thrust of their strategy for the world market was to gain control over the 50 percent of Russian exports "through Nobel."[30]

The press repeatedly published reports that Rockefeller owned a share of stock in the Nobel company. In early 1898 news reached Petersburg from M. Rice (serving in the Astor Business House, which was a Rockefeller affiliate) that the American trust "has direct and indirect interests in the Russian oil business."[31]

This was not the first such warning signal, and the tsarist government ordered its general consul in New York to make the necessary investigation. The consulate's manager, A. P. Veiner, reported that he had found that the board of Standard Oil, after careful examination of the situation in Baku, had resolved "to seize control of the oil business in Russia" in order to become "a producer inside the country itself." It had appropriated 10 million dollars for this purpose.[32] Approximately at this time Baku was visited by agents of Standard Oil, Suzard and Lufkin, who tested the waters: they examined the attitudes and plans of local oil producers and identified three firms for purchase, which would constitute the base for a unified production as a branch of the Rockefeller company.[33] Although the Baku oil men asked an exorbitant price, Standard Oil was prepared to negotiate. Nevertheless, nothing came of all this: the name of Rockefeller was simply too opprobrious for the tsarist government to approve such a deal.

As for the question of participating in the Russian oil business, American capital tried another approach. In the search for loans that the Ministry of Finance regularly had to undertake, in late 1898 Witte explored the possibility of obtaining credits in the United States. Although his representative came to the conclusion that "the amount of free money seeking a place for investment is hardly great,"[34] sometime later American banks offered to make Russia a loan. One of the first communications on this came from Rockefeller's representative, W. Libby. It is significant that the 10-million dollar sum that Veiner had mentioned earlier (assigned by Standard Oil for purchases to "take control of the Russian oil business") coincides with the amount of the American loan, which, after short negotiations, was in fact made at the end of 1899.

Negotiations for the loan were conducted by the bank of J. P. Morgan. The agreement was enveloped in the usual precautionary measures and did not provoke any suspicions that Standard Oil might be involved. The agreement was made in Morgan's name by New York Life Insurance, well-known for its activities in Russia. Business circles regarded Morgan and Rockefeller as rivals, which they doubtlessly were. Nevertheless, on this occasion the Morgan company actually acted as an agent for Rockefeller.

Behind the Morgan insurance company was a powerful consortium headed by the National City Bank of New York, which was called the Standard Oil Company Bank and whose director's name (James Stillman) was regarded as "a synonym for the name of Rockefeller."[35] In the consortium for floating the Russian loan, the bank represented the interests of the American oil trust. The main part of the loan (6.8 million dollars) went to finance the most important oil transport line in Russia—the Vladikavkaz Railroad. This alone was, in the words of the American historian L. Corey, "indicative of the widely cast international interests of the Standard Oil Co."[36]

The loan agreement of 1899 was followed by negotiations between National City Bank and the St. Petersburg International Commercial Bank regarding an expansion of Russo-American economic relations along various lines. As a result of Rothstein's visit to New York in June 1900, he reached an agreement with Stillman of City Bank to expand business ties between the Russian and American banks. "I am coming to an agreement with National City Bank about a deal for a joint account with the Russo-Chinese and International Bank," wrote Rothstein, summing up the results of his negotiations in the United States.[37]

The fact that Rothstein also represented the Russo-Chinese Bank in the negotiations was no accident. The fundamental reason was that, among the problems under discussion, a significant place was accorded to the question of American trade in the Far East, the fate of which evoked keen interest at Standard Oil.

In the mid-1890s the syndicate formed under Rockefeller's aegis and composed of representatives from the leading corporations on Wall Street, proposed to construct a Trans-Chinese railroad to support the development of American trade in China. The scheme collapsed, however, because of the fierce opposition of European powers.[38] The failure also coincided with losses on the commercial front. Acting through a nominal figurehead of Chinese nationality, Standard oil tried to create a peculiar sphere of influence in the densely settled province of Kwangsi [Guangxi].

Standard Oil succeeded in signing a contract with the provincial governor to establish a monopoly for the sale of American kerosene, but the agreement was nullified under the pressure of the great powers.[39]

At the end of the 1890s the American oil trust encountered mounting difficulties on the Far Eastern market. Apart from the factors linked to the struggle among oil corporations, the American interests were further threatened by the division of China into spheres of influence for European powers and Japan. The United States itself would have liked a piece of the Chinese pie, but was not properly prepared to compete with the European powers.

Amidst this division of China into spheres of influence, the United States launched a campaign to observe the principle of equal opportunities in China. Under the leadership of a Standard Oil representative (J. McGee), a Committee on American Interests (later reconstituted as the American-Asiatic Association) was formed. Representatives of American business circles appealed to the U.S. government to take vigorous measures to support American interests.[40] Under the pressure of Wall Street, the State Department demanded that European powers and Japan observe the principle of open doors.

The heart of the open-door doctrine was the demand by the United States that American commercial firms have equal opportunities with companies of other powers. That is to say, the doors should be open everywhere and to everyone, independent of any spheres of influence. Given the economic predominance that the United States had achieved by that time, it is clear that open doors would put them in an advantageous position.

A key element of the Russo-American Bank agreement was the attempt to free America's Far Eastern trade from dependence upon the British. Even in the early twentieth century, the financing of American exports to the East, as a rule, was still obtained through English banks. Many American goods, including petroleum products, were shipped through London. A significant part of the shipping was carried by English vessels, and the American goods in China itself were sold by branches of British firms. Even Standard Oil, which had built a colossal fleet of tankers and had agencies in a number of Chinese ports, was constantly forced to use the services of British capital and British commercial representatives, in particular, Jardin Mathieson and Company.

The agreement between Stillman and Rothstein envisioned reorganizing the system of financing American trade in the East, essentially with the aim of severing the dependence upon Great Britain. "In general it is our goal," wrote Stillman in his summation of the negotiations, "to divert from London ... the financing of exports from American to China and East Asia" in order to be able to conduct these "directly." In practical terms, this meant that calculations previously made by English banks in London would now be made on location by the Shanghai Branch of the Russo-Chinese Bank. These operations were to be performed by "each party" providing "cover for its own proportions" or with the funds of the National Bank crediting the Russo-Chinese Bank "at a moderate rate."[41]

Thus, the idea was to create a Russo-American banking association to operate in the Far East, with its primary goal being to promote the success of Standard Oil. In any event, from the list of firms that (in Stillman's view) would organize their commercial transactions through the Shanghai Branch of the Russo-Chinese Bank, approximately half were companies engaged in the sale of American kerosene.

This approach did not hold the slightest interest for the Russian side, whose main goal was simply to obtain credits. To make matters worse, circumstances became unpropitious for the Russian aspiration: the American stock exchange registered a sharp downward turn. That bearish tendency was intensified by the Chinese Boxer Rebellion in late 1900: although a united military expedition suppressed the rebellion, disagreements erupted among the great powers over the policies that should now be pursued in China. These events also affected Russo-American relations. J. P. Morgan, who had participated in the negotiations with Rothstein, categorically refused to continue them. In a telegram to Rothstein he wrote: "Having given the matter careful consideration on our part it will be absolutely impossible in our opinion to do anything until Chinese question satisfactorily settled. Morgan."[42]

When that telegram was put on Witte's desk, he immediately understood that Morgan refused to float the Russian loans. An infuriated Finance Minister gave vent to his rage in a resolution inscribed on the Morgan telegram:

> Please ask Rothstein to appear and inform him from me that I ask him never, without my permission in each case, to have relations with any foreign bankers on questions involving the monetary operations of the Russian government, since these operations concern this government—i.e., all direct and indirect loan operations. Explain to him that such dealings lead foreign bankers to assume that the government needs money and seeks loans—which substantially harms the state's credit standing.

Witte of course knew perfectly well about the negotiations with Morgan and simply decided to make Rothstein the scapegoat. Indeed, the negotiations with the Americans continued, although they did not actually achieve any progress. What the Russian side sought proved unattainable; what the American sought was unacceptable to the Russians.

In the interim Rockefeller's representatives continued to put on pressure, under the assumption that—sooner or later—they would achieve their goal. At the end of 1901 Standard Oil asked M. V. Rutkovskii (an agent of the Ministry of Finances stationed in Washington) to obtain on its behalf permission from the tsarist government to purchase oil fields in Baku. Several days later the firm made the same request of Stillman, supplementing its query with an unveiled threat. It told Rutkovskii that, if Russia does not agree to admit American capital to the oil fields around Baku, Standard Oil will offer a loan to Japan, which would be manifestly be used to prepare for war. As Rutkovskii reported, "Messrs. Rockefeller promised to offer 12.5 million, as Stillman told me, because the Japanese government responded favorably to the opening of Standard Oil's operations in Japan."[43]

Stillman's communication bore the unmistakable tone of an ultimatum. That impelled Rutkovskii to telegraph St. Petersburg for instructions on how he should respond: "What should I reply to the Standard Oil Co. regarding the proposal, which was presented in [my] report of 1 January? The reply will make it easier for me to act against the Japanese loan that has been announced here."[44] Rutkovskii underscored this point in a letter to Witte, noting that "it would be possible to hinder [the issue of a loan to Japan], if I could inform them [the Rockefellers] of Your Excellency's favorably reply to their petition to grant Standard Oil access to the extraction, processing and shipping of Russian oil... In the latter case I would have grounds to counsel against the acceptance of Japanese bonds, the purpose of which is to receive the means for war against Russia."[45] At first Witte did not reply, but when he did telegraph New York, he merely declared that, for the present, he cannot give "any answer" with respect to Standard Oil.[46]

John F. Archbold, the son of a boss in the Rockefeller trust, arrived in Russia in the summer of 1902. The visit had been arranged earlier with Nobel, and apparently the American company counted on the support of the Nobel firm. The latter organized a welcoming reception for the Rockefeller emissary. Nobel also gave an instruction to the firm's Baku branch to render Archbold every possible assistance. Informing the branch of the impending arrival of "the son of the chief representative of the American company, Standard Oil Co., who is now making a journey to become familiar with the oil business," Nobel asked that Rockefeller's emissary be given "special attention." "Suggest to him that he stay at our villa," ordered Nobel, "and put at his disposal the quarters in the director's house." Nobel emphasized that, "in view of the friendly relations of our board's directors to persons at the head of this great American business," it is necessary to give Archbold complete assistance so that his stay in Baku "leave the most pleasant memory" for having "given the young traveller the possibility to combine pleasure with work."[47]

It is not known whether Archbold derived pleasure from his journey to Russia, but so far as the work is concerned (in the sense that the representatives of Standard Oil understood this), this time they did not have good fortune. As the American historians Ralph and Muriel Hidy note, young Archbold was joining the Company's management at that time and seeking a means to bolster the eroding position of Standard Oil.[48] Nobel probably knew about this, and his willingness to assist Rockefeller's emissary hardly went so far as the Americans wanted, all the more since the tsarist government remained intractable. As earlier, it was hostile to Standard Oil and rejected Rockefeller's claims to the oil in Baku—above all, because that was contradictory to the interests of Russian companies as well as the circles of foreign capital, whose interests the tsarist government was forced to take into account. Although Rothstein's agreement with National City Bank remained in force, its real effect was far more modest than the American side had envisioned: only on a negligible scale did the Shanghai Branch of the Russo-Chinese Bank undertake to conduct American export operations and none of these involved the sale of American oil.

All this inevitably affected the development of Russo-American relations. If in the winter of 1901-1902, when Standard Oil had not abandoned hope of obtaining oil fields in Russia, Rockefeller refused to participate in a Japanese loan, afterwards National City Bank took a leading role in financing Japanese preparations for war.

FORMATION OF THE ROYAL DUTCH SHELL TRUST: THE PURCHASE OF ENTERPRISES IN RUMANIA, RUSSIA, AND AMERICA

Rockefeller's ties with Nobel, the American attempts to gain entry to oil production in Russia, and the creation of the Russo-American Bank—all inevitably evoked great anxiety among Standard Oil's rivals in the Far East.

The Russo-Japanese War of 1904-1905 intensified the concern over the activities of the American oil trust. Although the United States declared its neutrality, in fact its policy tended to be more favorable toward Japan.[49] The reason for this was the fact that American capital had a vested interest in a weaker Russia. One of the main consequences of the war was a sharp increase in American oil exports to the Far East, which shifted the alignment of forces in favor of Standard Oil, and the latter began to squeeze out not only the Russian oil exporters, but also Deterding's Royal Dutch company. Rockefeller's representatives said openly that the Russo-Japanese War had played right into their hands. "We were faced with a difficult task," they confessed, "and if we were successful in solving it, it is only because of the fact that Russia was drawn into war with Japan."[50]

The Russo-Japanese War became a landmark in the development of international rivalry, for it changed the alignment of forces and called new blocs into existence. Perhaps most important of all, it led to the merger of Royal Dutch Co. and the British Shell Co. Taking advantage of the latter's financial difficulties, Deterding proposed to Samuel a merger of the two firms, and in

February 1907 the two signed an agreement for the final merger of Royal Dutch and Shell. Thus began the existence of the oil trust "Royal Dutch Shell." The terms of merger guaranteed Deterding's predominance, for Royal Dutch held 60 percent of the stock and Shell only 40 percent.

A number of factors lay behind this all-important merger. The increasing competition in international oil demanded that the two partners of Royal Dutch Shell combine efforts against Standard Oil, which in 1906-1907 launched yet another round of price wars. Still another factor was the appearance of a new and dangerous competitor—German banks. After entering the oil industry in Rumania and demonstrating initiative in the creation of a syndicate to control the petroleum business in Europe, the German banks made a serious claim to participate in the petroleum club of leading producers. Falling petroleum prices on the international market also played a role, a direct consequence of the strong pressure from the new American oil companies, which were outside the control of Standard Oil and each year increased their share of the petroleum trade.

That glut on the market reached sizable proportions. The increases in output were enormous; for the period 1902-1907, for example, world output rose 150 percent. Much of that increase was due to the 200 percent increase in North America and also to the rapid development of the oil business in new areas (in particular, in Rumania). The Dutch Indies showed no significant change, while oil production in Russia actually declined.[51] Such changes, of course, were bound to have a profound impact on the alignment of forces. Compared with the first years of the century, when Royal Dutch and Shell first combined their efforts, their share of oil products had significantly declined by 1907. The volume of commercial operations by the concern "Asiatic" did not suffer, but the latter was forced to purchase enormous shares of oil from companies not belonging to Royal Dutch Shell. To correct the situation, Deterding decided to take measures that aimed to broaden the production base of the Anglo-Dutch trust.

He began with the 1908 purchase of oil fields in Rumania, which by this time had become an important oil base in Europe. Following the Germans and Americans, Deterding established his own company there—called Astra-Romana. He planned to monopolize the Rumanian oil business and tried to draw into the orbit of Royal Dutch Shell the largest foreign company there—the German firm Steaua Romana. Although nothing came of this effort, Deterding's Astra-Romana did succeed in overtaking the Germans and attaining first place among the foreign firms operating in Rumania.

Almost simultaneously, Deterding focused on Russia, where the appearance of Royal Dutch Shell coincided with a new wave of attack by British capital on the Russian petroleum industry. As we will see, the negotiations that Deterding began with the Rothschilds, culminating in one of the largest oil deals of the time, which permitted Royal Dutch Shell to establish a base in Russia.

No less important for the fate of the Anglo-Dutch trust was the fact that Deterding also established a base in America. For a long time, Rockefeller had loomed as the unchallenged master of the Western hemisphere. But this could not last forever. In 1907, in response to a price war that the Americans had unleashed in Europe, Deterding ordered his tankers, which had been heading for German ports, to change course and unload in New York. It was an act of unprecedented impertinence that no one else had ever allowed himself; it was this operation that earned him fame as "the Napoleon of oil."

"Until we started trading in America," Deterding later wrote, "our American competitors controlled world prices, because ... they could always charge up their losses in underselling us in other countries against business at home, where they had a monopoly." In order to "put an end to this state of things," the leaders of Royal Dutch Shell decided that "America, with her vast resources both for production and trading" should be included "into our general working plan."[52]

An attempt, however, was made beforehand to come to an amicable agreement with Rockefeller, and with this goal in mind Deterding visited New York in 1907. But the terms he proposed were rejected by the Americans, and the two sides could not reach an agreement. Many years later, in an attempt to justify his expansionist actions, Deterding wrote: "We had no other alternative but to expand, expand, expand." As Deterding explained: "Had we restricted our trading purely within certain areas, our competitors could easily have smashed us by relying on the profits they were making in our countries to undercut us in price. So, to hold our own, we had to invade other countries too."[53]

In 1907, shortly after the sudden appearance of Royal Dutch Shell's tankers at the port of New York, Deterding made his first attempt to establish a base in American oil production. But this first attempt ended in failure. It was not until several years later that the leadership of Royal Dutch Shell managed to implement its plan to gain holdings in the United States.

The Anglo-Dutch trust endeavored to exploit the emergence of rivals to Rockefeller in Texas, Oklahoma, and Indiana, where large new oil fields were discovered shortly after the turn of the century. The newly discovered oil fields provided the basis for such powerful firms as Texas Oil Co., Gulf Refining Co., etc., and as a result the share of the Rockefeller trust declined noticeably. If, at the beginning of the century, Standard Oil controlled approximately 95 percent of the oil industry in the United States and 90 percent of the oil exports, by 1910-1912 this situation had undergone dramatic changes: about one third of the oil production now belonged to "independent companies, which spurned attempts by Standard Oil to make agreements and instead struck a deal with Rockefeller's arch-rival (the European Oil Union) to deliver large quantities of kerosene and gasoline each year.

In late 1910 and early 1911, Deterding opened negotiations with Rockefeller's American competitors. He made it his goal to create an anti-Standard Oil coalition in America, which could successfully compete with the Rockefeller company not only on foreign markets, but also within the United States itself. The negotiations lasted more than two years, but did not end in success. Deterding then shifted his gaze to the American West—to California. It was precisely at this point that large new oil reserves were discovered and became the focus of a new speculative fever. Standard Oil had already opened a branch there in 1900. Deterding decided to torpedo Rockefeller's California firm and, beginning in the fall of 1912, began dumping oil from the Dutch Indies at cheap prices. When it became clear that dumping alone was insufficient, Royal Dutch Shell attempted to acquire some property, but it was not until 1914 that Deterding established his branch in America (Shell Co. of California), which, over the years, became one of the largest oil corporations of America after absorbing a series of companies in Oklahoma and other states.

When Deterding decided to enter the oil industry of America, he also took into account the fact that Standard Oil was faced with an unfavorable situation because of the legal battles over its violation of the 1890 anti-trust law (prohibiting monopolistic organizations). To be sure, the leader of Royal Dutch Shell was not inclined to exaggerate the significance of this factor. In contrast to predictions in the press, Deterding never seriously believed that the judicial proceedings against the Rockefeller trust would end in effective sanctions.

In 1911, when the U.S. Supreme Court issued its decision, Standard Oil was the largest multinational corporation in the world, with stocks worth 660 million dollars and with 67 branches in the United States and abroad.[54] The Supreme Court ordered that the trust be dismantled. Nevertheless, even after this order, the branches continued to collaborate. The coordination of foreign operations, to a significant degree, was orchestrated by Standard Oil of New Jersey, which continued an active assault on the world market—a fact that has inspired "company historians" to label the period after 1911 the "years of resurgence."[55] For rivals of Standard Oil, the Supreme Court's decision offered little consolation, especially with respect to the trust's foreign operations, which the American government supported with all the means at its disposal. Nevertheless, the Supreme Court decision created more favorable conditions for the formation of new companies independent of Rockefeller. And Deterding took advantage of this.

After establishing bases in Rumania, Russia, and America, Deterding had substantially strengthened his position. And the men at 26 Broadway understood fully what the formation of Royal Dutch Shell meant and what consequences would ensue from an expansion in its sphere of influence. Later, in recalling these events, Rockefeller was generous with verbal abuse and vituperation for Deterding and his consorting with the Rothschilds. "I am not one," he said, "who would be willing to lie supinely on my back and have our dear English friends—much as we love them!—arrange our business affairs and bind us hand and foot. We are having too much experience today of their methods: taking our ships, using them for the benefit of our Asiatic competitors controlled by Jewish men who cry 'Wolf! Wolf! Standard Oil Company!'"[56]

Formation of Royal Dutch Shell and the expansion of its sphere of activity had the effect of making its competition with Standard Oil still more acute. The oil wars seized new regions of the globe.

NOTES AND REFERENCES

1. *U.S. Mineral Resources* (1899), p. 281.
2. Ibid.
3. F. C. Gerretson, *History of the Royal Dutch* (Leiden, 1958), *2*, p. 351; *U.S. Mineral Resources* (1899), p. 259.
4. Gerretson, *1*, p. 282.
5. Ibid., *1*, 282-283.
6. R. Hidy and M. Hidy, *Pioneering in Big Business* (New York, 1955), p. 263.
7. Ibid.; Gerretson, 2, pp. 46-48, 57.
8. R. and M. Hidy, pp. 264-265.
9. Gerretson, *2*, p. 61, 63.
10. Report of the U.S. Consul in Batavia, S. Everett (dated March 18, 1898). *U.S. Consular Reports*, 57, p. 380.
11. Memorandum of the General Consul of Russia in Tientsin, Shuiskii (April 5/17, 1898). In *AVPR*, f. Kitaiskii stol, d. 1219, l. 15.
12. Report of the U.S. Consul in Batavia, S. Everett (November 31, 1899). In *U.S. Commercial Relations* (1899) *1*, p. 882.
13. Gerretson, *2*, p. 72.
14. Memorandum from Shuiskii (April 5/17, 1898). In *AVPR*, f. Kitaiskii stol, d. 1219, l. 15.
15. Report of the British consul in Amsterdam (Robbins) for 1897. In *British Diplomatic and Consular Report*, A.S., N 2054, p. 7.
16. Report from Everett (October 31, 1899). In *U.S. Commercial Relations* (1899) *1*, pp. 881-882.
17. Report from Everett (dated February 1, 1899). In *U.S. Consular Reports*, 57, p. 55.
18. Vert to the Board of the Russo-Chinese Bank (February 15/27, 1898). In *TsGIA SSSR*, f. 560, op. 28, d. 538, l. 112.
19. M. M. Bakunin, *Tropicheskaia Gollandiia* (St. Petersburg, 1902), p. 337.
20. "Torgovlia russkim kerosinom v Kitae i na Dal'nem Vostoke. Spravka Shankhaiskogo otdeleniia Russko-Kitaiskogo banka" (March 20/23, 1900). In *TsGIA SSSR*, f. 560, op. 2, d. 188, ll. 37-38.
21. Report from Everett (October 31, 1899). In *U.S. Commercial Relations*. 1899, *1*, p. 882.
22. Ibid., *1*, p. 881; reports of the U.S. consul (B. Rarden) in Batavia in November 1900 and June 1901. In *U.S. Commercial Relations* 1901, *1*, p. 817; 1900, *1*, p. 972.
23. Gerretson, *2*, p. 101.
24. Ibid., *2*, p. 120.
25. H. Deterding, *An International Oil Man* (London, 1934), p. 68.
26. F. Lane to V. N. Kokovtsov (May 17, 1898). In *TsGIA SSSR*, f. 20, op. 7, d. 179, l. 172.
27. Deterding to Lane (June 19, 1902). In Gerretson, *2*, p. 238.
28. Hewins, p. 49.

29. *Moniteur,* October 5/18, 1902.
30. Transcript of a conversation between Rockefeller and Inglis (November 14, 1918) *RAC.*
31. M. Rice to Kovalevskii (January 26/February 8, 1898). In *TsGIA SSSR,* f. 20, op. 7, d. 179, l. 5.
32. Report from Veiner (May 13/25, 1898). In *MKNPR,* pp. 218-219.
33. R. Hidy and M. Hidy, p. 510.
34. Memorandum about the results of negotiations regarding the possibility of concluding state loans in the United States (December 19/31, 1989), sent from Witte to Lamzdorf (December 24, 1898/January 5, 1899). In *AVPR,* f. Posol'stvo v Vashingtone, d. 142, ll. 49-50, 47.
35. F. Landberg, *60 semeistv Ameriki* (Moscow, 1948), p. 71.
36. L. Korei, *Dom Morganov* (Moscow-Leningrad, 1933), p. 176.
37. Memorandum by Rothstein (no earlier than June 1900). In *TsGIA SSSR,* f. 626, op. 1, d. 1372, l. 50.
38. See A. A. Fursenko, *Bor'ba za razdel Kitaia i amerikanskaia doktrina "otkrytykh dverei"* (Moscow-Leningrad, 1956), pp. 42-63.
39. Ibid., pp. 77-78.
40. C. S. Campbell, *Special Business Interests and the Open Door Policy* (New Haven, 1951), pp. 35-55.
41. J. Stillman to Rothstein (July 12, 1900). In *TsGIA SSSR,* f. 626, op. 1, d. 253, l. 13.
42. Telegram from Morgan to Rothstein (November 28, 1900), with Witte's resolution. In *TsGIA SSSR,* f. 560, op. 28, d. 50, l. 66.
43. M. V. Rutkovskii to Witte (January 13/26, 1902). In *TsGIA SSSR,* f. 560, op. 22, d. 249, ll. 136-137.
44. Telegram from Rutkovskii (not earlier than February 10/23, 1903) in ibid., l. 151.
45. Rutkovskii to Witte (February 12/25, 1902). In ibid., l. 156.
46. Witte to Rutkovskii (February 27/March 12, 1902). In ibid., l. 58.
47. Nobel to I. G. Garsoeff [Garsoev] (June 25/August 8, 1902). In *TsGIA SSSR,* f. 798, op. 1, d. 158, l. 72.
48. R. Hidy and M. Hidy, pp. 500-501.
49. See, A. Dobrov, *Dalnevosotchnaia politika SShA v period russko-iaponskoi voiny* (Moscow, 1952), chapters 2 and 4; R. D. Challener, *Admirals, Generals and American Foreign Policy, 1898-1914* (Princeton, 1973), pp. 222-223.
50. *Neftianoe delo,* November 15/28, 1905 and December 15/28, 1905.
51. See Table 5 in the Appendix.
52. H. Deterding, *An International Oil Man* (London, 1934), pp. 87-88.
53. Ibid., p. 88.
54. R. Hidy and M. Hidy, p. 637; Wilkins, p. 84.
55. G.S. Gibb and E.H. Knowlton *The Resurgent Years, 1911-1927* (New York, 1956).
56. Transcript of a conversation between J. D. Rockefeller and G. D. Inglis (May 29 and November 14, 1918), *RAC.*

Chapter IV

The European Petroleum Union and the "Kerosene Comedy" in Germany

The battle over oil led, from time to time, to the formation of alliances, which served to confirm a realignment of forces among corporations and financial-industrial groups. Thus appeared the European Petroleum Union, established at the initiative of German banks with the purpose of securing for Germany a position in the petroleum club equal to that of the other powers. By 1907 the twenty million Marks originally invested in the European Petroleum Union had increased to 37 million marks, of which 20 million belonged to the German banks.

THE FORMATION OF AN ANTI-AMERICAN BLOC IN EUROPE

The abortive attempt by German capital in the late 1890s to establish a base in Russia did not end their attempts to create an anti-American coalition. Having lost hope of purchasing the Baku oil fields, the German banks turned their attention to Rumania, which, after the turn of the century, was the site of new oil discoveries that provoked a speculative fever and stirred German hopes of acquiring its own oil base.

In 1902 the petroleum press observed that Rumania had become "the object of special attention on the part of competent German circles," which had come to the conclusion that the Rumanian oil reserves "represent a special interest for Germany." Therefore the Berlin bank of Bleichröder, the affiliated "Disconto Gesellschaft" and the Deutsche Bank (later the key figure in German oil) concentrated their attention on Rumania.

After making an agreement with the Parisian Rothschilds and the Shell Co. of England, the German banks formed, in 1902, a company to manage the kerosene trade in Germany and to oppose Standard Oil. A year later this company concluded a contract for oil deliveries with the Rumanian firm Steaua Romana, which by then had been acquired by the Deutsche Bank. German planners ascribed a great importance to Rumania. The head of the Deutsche Bank, A. von Gwinner, admitted outright that Germany aspired to a "leading role" in the Rumanian oil industry and sought to gain control of Steaua Romana, at the time the largest company for the production and sale of Rumanian oil. After increasing its investment from 10 to 17 million lei, then to 30 million lei in 1906, the Disconto Gesellschaft and Bleichröder were content with a smaller role after they had bought up several smaller firms. Nevertheless, by 1906 the aggregate capital invested by German banks in Rumanian oil companies amounted to 97 million lei—64.3 percent of all foreign investment in the Rumanian oil business.[1]

The German banks acted upon the initiative and with the active support of their government. The highest imperial circles wanted Germany to become an independent oil-producing power. The Deutsche Bank's campaign in Rumania, for instance, immediately followed a special discussion in 1900 in the Ministry of Foreign Affairs. When Steaua Romana fell into German hands, Chancellor von Bülow sent Gwinner a letter, in which he noted "with satisfaction" that he had observed the "energy of the Deutsche Bank," which had "succeeded in participating in a foreign enterprise," and he "decisively" welcomed its activities for providing "the domestic market of the entire German economy with petroleum."[2]

During debates in the German Reichstag, deputies demanded measures against the Rockefeller yoke. Rumania was clearly regarded as a launching pad for the attack on Standard Oil. The diplomatic service of the United States noted that "a powerful combine, composed of leading German banking institutions—among them the 'Deutsche Bank' and the firm S. Bleichröder in

Berlin—has formed close connections with Romanian oil companies, having large territories, now to be worked by means of the capitalistic power furnished by the German banks."[3] The German press unleashed an anti-American campaign. At its own expense, the Deutsche Bank published a symbolically significant book—*The Thirty Years Oil War,* whose authors depicted victorious prospects of future skirmishes with the Rockefeller Trust.[4]

These cries of victory, however, were soon drowned out by an American counter-attack. In Rumania, where everything had seemingly proceeded so favorably for German interests, Rockefeller created his own branch (Romana-Americana) in 1904 as a counterweight, with a relatively modest investment capital. At first the capital of the American firm in Rumania amounted to just 2.5 million lei, which was later increased to 5 million lei, but that still represented a mere 3.3 percent of total investment in the Rumanian oil business. Immediately after this the Americans constructed two oil refineries in Austria-Hungary, thereby expanding their base in Western Europe. That action was fraught with still greater danger for the Germans, who had established themselves earlier in the Galician oil fields. The Germans planned to form a triadic German-Austrian-Rumanian oil bloc against Standard Oil, and the American representatives were not wrong in judging the whole operation as an effort to destroy Rockefeller's commercial branch in Germany.

The escalation of the German-American conflict stiffened the German resolve to create an organization capable of resisting the American oil trust. As early as the summer of 1904, the Russian paper *Neftianoe delo* noted, that "nowhere is the effort to be rid of Standard Oil's dominance of the market so strong as in Germany, and nowhere has this question been discussed with such fervor, not only in the general press, but even within the walls of the Reichstag, where it has become intertwined with questions of general state importance," to the degree which has happened in Germany.[5]

Indeed, the conflict with the Rockefeller oil trust developed under conditions of general struggle between Germany and the United States for dominance in world industry and trade. The struggle cast long shadows not only over the bilateral relations of the two countries and also the larger framework of international relations. S. von Waltershausen, a professor of political economy hired by the German government to write a brochure on these questions, noted that the United States, "with increasing growth of its economic power" has gained in "political might" as well. He therefore renewed the call for a commercial and political coalition specifically directed against the Americans.[6]

The German press continued to sound the alarm with regard to "the American danger," and this anti-American clamor found a sympathetic response in official circles. The German ambassador in Washington, Holleben, urged that, "there is still time to confront America directly" in order to prevent a "flood of American exports" and to stop American capital from "swallowing up European enterprises." His opinion was shared in the Berlin ministries.[7]

Business circles hastened to confirm such views. In 1904 the Deutsche Bank, along with its allies, created a conglomerate called the Deutsche Petroleum Aktiengesellschaft, with a capital of 20 million marks. This holding company assumed the responsibility for all the oil firms of the Deutsche Bank in Rumania, Austria-Hungary and Germany itself. The controlling bloc of stocks was held by the Deutsche Bank. The Disconto Gesellschaft, Bleichröder, and allied banks established a similar kind of conglomerate, the Allgemeine Petroleum Industrie Aktiengesellschaft, with a capital of 17 million Marks in 1905. This was also a holding company, coordinating the activity of their oil enterprises in Rumania and Germany. The two holding companies were independent of each other, but maintained close business contacts.[8]

The growth of anti-American sentiments in Germany were used by the Deutsche Bank to form the European Petroleum Union, which was virtually a replicate of the German proposal of the 1890s for a petroleum coalition in Europe. From time to time, Berlin renewed its interest in combining the efforts of German capital's new European oil firms with the forces of Russian oil producers. A similar effort had been undertaken in miniature in 1901, when German capital achieved joint participation with the Rothschilds in Steaua Romana. The resulting agreement was insignificant in scale, but circumstances had subsequently changed, impelling Russian oil producers to take greater interest in contacts with Germany.

What united the two sides, above all, was the waxing aggressiveness of Standard Oil. Moreover, by the time the European Petroleum Union was formed in 1906, certain changes in the status of the two parties helped to facilitate the agreement. On the one hand, Russian oil magnates were now encountering serious difficulties because of the sharp fall in the production and export of kerosene in the wake of Russia's defeat in the Russo-Japanese War and the outbreak of revolution in 1905.[9] On the other hand, once Germany had acquired its own oil fields in Europe (chiefly in Rumania), to some degree it had strengthened its position in the impending negotiations with Russia. German capital expected to reach an agreement with the Russian oil magnates on term that would give them the deciding voice in the planned union.

The creation of the European Petroleum Union under the aegis of the Deutsche Bank was significant in its own right, for it was precisely this bank that stood in the avant-garde of German expansion and took the initiative in many of Germany's international operations. The political dimension of the oil question was further complicated by the naval rivalry—that is, by the prospect of creating faster ships using improved crude oil engines. The issue acquired new urgency after J. Fischer's appointment in 1904 as lord of the British Admiralty, for he was renowned for his obsession with the idea of converting the British navy to oil fuel. As American intelligence observed, Germany now began to show growing interest in a "petroleum rearmament" of the fleet, although in this area the Germans lagged far behind the English.[10]

Germany thus was hastily seeking to become a great oil power and displayed an "enormous interest" in controlling the production and trade in oil products. According to data in the magazine *Moniteur* (which was well-informed on the oil business), the European Petroleum Union was formed "under the direct supervision of the German government." In explaining this interest of the German authorities, the magazine observed that it was not simply a matter of seeking to spite the Americans, but "a concern about providing the German navy with deliveries of liquid fuel," which appeared to be "the fuel of the future." Guided by these goals, the German government "decided to encourage German capital to participate in the petroleum business."[11] To be sure, the European Oil Union was created as a marketing corporation for the trade in kerosene and diesel oil, whereas in the case of lubricants, gasoline and other liquid fuels, the agreement contained the proviso that participants of the Union "retain their complete freedom everywhere." But the German government's point of departure was "the simple idea" that liquid fuel exists wherever there is kerosene. It was no accident that a report of the Deutsche Bank (devoted to an analysis of the development of industrial technology since the turn of the century) noted the growing importance of oil as a fuel for diesel and other motors and emphasized the need for Germany to be guaranteed oil deliveries.[12] Although kerosene still held first place in Germany's consumption of petroleum products (approximately 1,000,000 tons per annum from 1901 to 1911), the import of liquid fuel had increased by a factor of 42, rising from 6,000 tons in 1901 to 250,000 tones in 1911.[13]

In the opinion of interested business and political circles in Germany, the most important fact was the creation of the European Petroleum Union as the foundation for a European oil trust. It was said that the Union's formation served to unify the interests of both the Deutsche Bank and the banks of Bleichröder and Disconto Gesellschaft, based on their common desire to conclude a pact with the largest Russian oil firm—the Nobel Brothers Company. Although only agents of the Deutsche Bank signed the contract with the Nobel-Rothschilds group to create the European Petroleum Union, the other banks still took an interest in Russian oil. Thus, noting that it had assisted in increasing Nobel's capital and had placed its stock on the Berlin Stock Exchange, in early 1907 the Disconto Gesellschaft expressed its willingness to render financial assistance once more.

It would be an exaggeration, however, to suggest that the German banks were fully unified and in total agreement in petroleum dealings. To be sure, the Disconto Gesellschaft did make the offer of financial assistance for the Nobel firm, and the latter hastened to seize this opportunity. But Nobel also exploited the internal divisions of the German banks to conduct competitive negotiations with each separately. Revealingly, when the Deutsche Bank learned of the overtures from Disconto-Gesellschaft, it also expressed a willingness to give financial assistance to Nobel, but only on the grounds that this be independent of any agreement involving the Disconto bank. The head of Deutsche Bank, Gwinner, declared categorically that his bank "cannot be useful" to Nobel in a joint undertaking with "the two other local banks" of Disconto and Bleichröder. However, he did offer to render services "in any other form," including the purchase of old Nobel stock for up to a sum of 3 million rubles. Simultaneously, Gwinner expressed a readiness to participate in the plans to increase the capital of the Société anonyme d'Armament, d'Industrie et de Commerce (SAIC), an association for the trade in lubricants, in which Nobel had taken 2 million francs and held the dominant influence. The Deutsche Bank said outright that it had already endeavored for many years "to establish a connection between the most prominent and significant Russian corporation and our bank."[14] From what the Germans had concretely proposed for the realization of this goal, it was perfectly obvious that more than just kerosene was at stake. The German calculations ran much further, if one keeps in mind the entire broad circle of interests that enveloped the oil business. In addition, everything that the Deutsche Bank strove to obtain from Nobel it did so for its own sake, and contrary to the Disconto Gesellschaft.

Just how unjustified was the optimism that the European Petroleum Union could unite the anti-American forces of Europe became especially apparent in the ensuing relations between the new association and Standard Oil. Striving with all its might to prove the thesis of unity among the German groups, the *Moniteur* refuted widespread rumors that the Disconto Gesellschaft had made a deal with Rockefeller. True, the journal was forced to admit that a cartel-like connection did exist between the oil producing firm of Disconto in Rumania and the Rockefeller branch of Romana Americana. But the journal emphasized that the German bank maintained an "independent" position and backed up this argument with an official statement from the Disconto Gesellschaft.[15] As portrayed by the *Moniteur,* the European Petroleum Union was a monolithic wall that slammed the door shut in the face of American penetration. Those who thought this way, however, were in for a bitter disappointment. In mid-1907 the European Petroleum Union signed a pact with Standard Oil to divide up the markets and to fix their respective quotas for kerosene, according to which the Americans secured the right to import 75 percent of the kerosene to Europe, with just 25 percent left for the Deutsche Bank and its partners.

According to the testimony of a Russian financial agent in Germany, P. Miller, who had information on the course of the negotiations, these talks were for a long time kept "in strict secrecy" and the very fact of their existence had still not been reported in the press by early 1907: "The Americans are acting in a restrained manner" and seek an agreement with the European Petroleum Union and "especially the Russian industrialists."[16] These negotiations were completed by mid-1907 with the signing of an agreement to allocate quotas for the oil trade in Europe.

How could it happen that an irreconcilable adversary—which is what Rockefeller had been to the Deutsche Bank—suddenly proved to be its ally? In the practice of corporations such a sudden volte-face was no rarity. An aggravation of differences gives way to a truce, which both sides use to regroup their forces and to prepare for new battles. To a certain degree, agreements always have a certain compulsory element, but this particular agreement was doubly so for the Deutsche Bank. It had to take this step, firstly, as a result of the price war begun by Standard Oil, which had inflicted great losses on the European Petroleum Union; secondly, because of the impossibility of securing sufficient oil without the American deliveries; and, thirdly, because of pressure from their own "allies," who had struck a deal with the Americans behind the back of the Deutsche Bank.

It turned out that, at the very start of negotiations to create the European Petroleum Union, the Deutsche Bank's partners were conducting parallel negotiations with Rockefeller's representatives. This task was again assigned to Lane, who, with his characteristic adroitness, prepared a plan for an agreement between the Deutsche Bank and the Russian firms, and began negotiations with the head of Standard Oil's English branch, James McDonald.[17]

Earlier, the Rothschilds had constantly accused Nobel of an excessive complaisance and tractability with respect to Rockefeller. And now their own representative, Lane, began to flirt with the Americans. It was later noted in one of the documents of the European Petroleum Union that "a bitter enmity exists between the Deutsche Bank and the Rothschilds."[18] It was explained by a general deterioration in the relations between France and Germany. However, the condition of the Russian oil industry in 1905 did not permit the Rothschilds to spurn the German proposal to organize a joint association. Still, the French had no intention now of becoming dependent on the Germans. And that is why the Rothschilds wanted the ties with the Americans, since that would exclude the possibility of dependence and a *diktat* from the Deutsche Bank. But if, under these conditions, the Rothschilds could not resist the temptation to establish ties with Rockefeller, then Nobel—which for many years had been bound by contractual ties with Americans—was prepared to make these relations still closer. Hence Nobel's representatives supported proposals to make an agreement with Standard Oil and participated together with Lane in the negotiations with Rockefeller's agents.

When the Deutsche Bank signed the agreement with Standard Oil, it did so because it had been presented with a *fait accompli*. Under the conditions of closer links between Rockefeller and the Russian oil industrialists, and the growing ties of the Disconto Gesellschaft with Americans, Gwinner simply had no alternative. Otherwise, it would have been left totally isolated. Nevertheless, the Deutsche Bank was a reluctant participant in the deal with Rockefeller and, literally from the first day of that the agreement was signed, it began preparations to wreck the arrangement. It was encouraged in this activity by the German government, which did not conceal its displeasure over the agreement. It was in these circumstances that circles close to the Deutsche Bank advanced a proposal in 1908 to establish a state monopoly on the oil trade in Germany. In the event it was approved, the agreement with Standard Oil (regarding oil deliveries to Germany) would immediately lose validity. Moreover, the proposal to establish an oil monopoly drew a favorable response in high circles and also in the German press. Nevertheless, it did not immediately prove possible to commence real discussions and attain legislative approval. Only three years later did the next fit of anti-American outbursts and unceasing attempts to solve Germany's oil needs lead to debates in the Reichstag.

In any event, the attempt to create a "single front" against the "American threat" to Europe was manifestly doomed to failure. The European Petroleum Union did not justify the expectations that had been nourished by German capital. Nor did the Russian side obtain the anticipated results; that was especially true of Nobel, which had hoped that the Union would strengthen its position against the growing competition with Standard Oil. During the three years that the Union existed, Nobel increased its export of kerosene to Europe three-fold: from 2,165 million poods in 1906 to 8,887 million poods in 1909. However, the prices on kerosene steadily fell: Nobel's kerosene deliveries to the Union brought 53 kop. per pood in 1906, but only 50 kop. in 1909. Meanwhile, the kerosene sold within Russia brought between 1.17 rub. and 1.35 rub. per pood. All this brought the Nobel firm to conclude that the European Petroleum Union "serves our interests poorly" and it was above all necessary to find ways "to increase our sales inside Russia."[18]

NEGOTIATIONS BETWEEN THE DEUTSCHE BANK AND DETERDING

In the final analysis, the failure of the European Petroleum Union was in large measure due to the fact that relations among its participants were unstable. In any event, they were far from making the organization so monolithic that could topple a giant like Standard Oil. An additional complicating factor was the increased activity of British capital after the formation of the Anglo-Dutch oil trust, Royal Dutch Shell, in 1907. The latter was an irreconcilable enemy of Standard Oil, a fact that ostensibly should have played into the hands of the Deutsche Bank. But the success of Royal Dutch Shell did not delight the Germans, who regarded the English firm just as dangerous an enemy as the American company. But it later came to pass that the Deutsche Bank and Royal Dutch Shell pooled their forces in the struggle against Standard Oil.

Gwinner and Deterding later held talks on concluding a large-scale deal. The Germans proposed to surrender their share in the European Petroleum Union to Royal Dutch Shell in exchange for compensation of a global character. These negotiations, held in July 1910, were caused by a series of important changes in international oil competition. One of them derived from the fact that oil reserves had been discovered in the Near East in 1908 and that an Anglo-Persian Oil Co. had been formed, thereby creating yet another arena in the struggle for oil and causing a shift in the main axis of business and competition. Most important of all, not only had a rich, new oil-bearing region been discovered, but geopolitically it lay in a region that was extremely sensitive to the competition of the great powers.

Having participated in the construction of the Baghdad Railway Line, the Deutsche Bank did not conceal the fact that, apart from strategic objectives, it pursued a concrete objective—to obtain oil concessions in an area next to Persia: Mesopotamia in the Ottoman Empire. Therefore, when Gwinner and Deterding began their negotiations, the Mesopotamian question was in the same stack as the question of the European Petroleum Union. It is no accident that, no sooner had the Rothschilds learned of these negotiations (through F. Lane, serving again as intermediary) than they announced their acute interest in solving this question. "You know that the Mesopotamian matter has always been of keen interest to my chiefs, and in their opinion this is now the time to move forward in this question," Aron wrote to Lane. He informed him that the Rothschilds "feel very favorably about the proposed combination" and, to promote its realization, were ready to subsidize Deterding with a sum of 6 to 7 million francs to buy the stocks of the European Petroleum Union.[20]

If one believes Aron's letters, then Deterding achieved an oral agreement with the Deutsche Bank that the latter generally surrender to Royal Dutch Shell "all the interests that it has in oil enterprises." However, the asking price of the German side was apparently too high. As Aron telegrammed Lane: "We are very surprised that the matter of the Deutsche Bank has stalled and that there is nothing definite regarding the negotiations."[21] But neither the Rothschilds nor anyone else was fated to learn anything new about all this. A year later, one English business connected with the European Petroleum Union attributed the breakdown of negotiations to the fact that Deterding would enter this organization in no other capacity than "as chief." "For the present moment," he wrote, "I understand the policy of Mr. Deterding is to leave matters where they stand in the expectation that at a later period, when our Berlin friends have been chastened by losses, he will be able to carry through his plan of acquiring their interest at less expense than at the moment is possible."[22] Is it possible to believe this judgment? How adequately does it assess the cause of the breakdown in negotiations between Deterding and the Deutsche Bank? It is difficult to answer these questions with any degree of certitude. Of course, the financial calculations noted by the author of the letter played their role. But it is also beyond doubt that success in concluding an agreement between the two parties was impeded by the growing antagonism between England and Germany.

THE PROPOSAL FOR AN OIL MONOPOLY IN GERMANY

Seeking a way out of this situation, in March 1911 the German Reichstag unanimously approved a resolution urging the government to create an organization under imperial control for the kerosene trade.[23] Thus, it posed the direct question of establishing a state monopoly for the kerosene trade. Realization of this goal would forsee the creation of a German joint-stock company with a capital of 40 million marks, which would be comprised of contributions from the large German banks. To that end, the company's governing board would include government representatives. Given his influence in high circles, Gwinner doubtlessly expected that this enterprise would be created under the aegis of his own Deutsche Bank.

Once the Reichstag endorsed in principle the idea of an oil monopoly, the government began to prepare the necessary draft legislation. As a memorandum from the state treasury attests, it "constantly conducted negotiations with the Deutsche Bank and the Disconto Gesellschaft."[24] The latter formed a new holding company called the Deutsche Erdöl Aktiengesellschaft in order to unite their oil enterprises in Germany, Rumania, and Austria-Hungary, including the Austro-Hungarian marketing company Österreichische-Ungarische Petroleum Produkte. Of fifteen members on the governing board of Deutsche Erdöl, ten were bank representatives. This union, however, did not survive very long. At the very time when the Disconto Gesellschaft inclined to keep the agreement with Standard Oil, the Deutsche Bank decided to annul it.

In early 1912 the Deutsche Bank managed to persuade its partners in the European Petroleum Union to sunder the agreement with Rockefeller for quotas in the kerosene trade. The head of Standard Oil's German branch, Heinrich Riedemann, protested this decision, but a German court upheld the legality of the action taken by the European Petroleum Union. "A kerosene comedy," as Lenin called these events,[25] was performed with the expectation of luring business and government circles to support the idea of a monopoly. Success in this adventure could only be secured if Germany succeeded, once it was free from the sway of the American trust, in finding other reliable sources to supply oil. Because the attempt to reach an agreement with Deterding had ended in failure, the main hope now rested with Russia, where the decisive word belonged to the Nobel and Rothschilds' companies—although here too, it was impossible to manage without Deterding, since he had acquired the Rothschilds' enterprises in early 1912.

The European Petroleum Union concluded contracts for kerosene deliveries with Standard Oil's rivals—the American companies of Pure Oil, Gulf Oil, and Texaco Oil. Gwinner attempted to guarantee these deliveries to Germany on a long-term basis. But Standard Oil offered higher prices to its American rivals and thus undercut Gwinner's efforts.[26] As for Rumania and Austro-Hungary, where German banks dominated, the scale of oil production was simply too small to satisfy the demand in Germany. In 1911-1913 oil output in Rumania and Austro-Hungary constituted about 3 million tons, less than half of which was exported. From various sources Germany could only obtain a total of some 20 million tons, whereas its annual demand for oil already exceeded 100 million tons.[27] Therefore, Germany's sole hope was to persuade Russian producers to fill the gap.

When, in late 1911 and early 1912, the question arose of increasing the capital of the Nobel company from 15 to 30 million rubles, the German banks eagerly expressed their desire to participate. In accordance with the ensuing agreement, Nobel stock was admitted to the Berlin Stock Exchange for 15 million rubles.[28] The new issue of stocks was carried out by the Disconto Gesellschaft, which had participated in an earlier distribution of securities for the Nobel firm. This time the bank retained for itself a bloc of stocks worth 6 million rubles. The operation of the Nobel stock was a profitable venture, and the Disconto Gesellschaft bank earned a substantial sum: the stock was initially taken at 180 percent, at the outset of the following year was quoted at 350 percent, and by September had reached a sensational rate—455 percent. By participating in this step to expand Nobel's capital, the German banks had certain expectations that they would now be able to persuade Nobel to support the kerosene monopoly. When the Nobel stocks were admitted for ratings on the Berlin Stock Exchange in April 1912, the German side simultaneously invited the firm to sign a long-term contract for oil deliveries to Germany. In doing so they referred to the fact that "several years ago you spoke ... of the possibility of the Nobel Brothers assuming a long-term obligation for deliveries."[29]

The Disconto Gesellschaft, like the Deutsche Bank, supported the kerosene monopoly, and undertook to negotiate with Nobel. But the latter declined to make an agreement, citing "the changed condition of the market."

Nevertheless, the Germans did not lose hope. Their determination sprang not only by the need for kerosene, but also from the growing interest in oil as a naval fuel, as the resident for the American naval intelligence continued to report.[30]

In early 1912 a memorandum of the Russian Naval General Staff emphasized that the question of introducing oil motors for naval vessels was being carefully studied by the German command and that "Emperor Wilhelm himself" was keenly interested in this.[31] Nevertheless, as the Russian naval agent in Berlin (Behrens) observed, right until the fall of 1912 the entire question of transferring the fleet to oil was "only in the initial stage." Oil burners had been installed on just 48 destroyers of a new type, but on the large vessels—the battleships and battle cruisers—neither pure oil nor a mixed heating system had been introduced. Even modernized ships were only partly adapted to use oil fuel. In those cases, when three boilers were installed on cruisers, only one used oil, the other two still used coal; or, where the vessel had four boilers, two used oil, the other two coal. "Destroyers that have this kind of heating system," wrote Behrens, "under ordinary circumstances sail under coal, and only during long voyages burn oil."[32] At the next session of the commission of the Reichstag that discussed the question of measures for the kerosene monopoly, the government's representative noted the growing significance for

Germany that it be provided with residual fuel oil and gasoline for military purposes.[33]

Considering these circumstances, the government of Germany decided to include the oil question on the agenda for a meeting between the German Kaiser Wilhelm II and the Russian Emperor Nicholas II in the summer of 1912. They hoped that negotiations at the highest level could solve this problem. The Kaiser did raise the question in a conversation with the chairman of the Council of Ministers, V. N. Kokovtsov. Emphasizing the "exceptional significance" which "in our time one should give to liquid petroleum fuel for both commercial and naval shipbuilding technology," Wilhelm proposed (according to Kokovtsov) to create a pan-European oil bloc. In essence, the Kaiser refloated the old idea of a commerical and political coalition against the United States, which he had proposed to the tsarist government during his visit to St. Petersburg in 1897. But on the first occasion, neither the proposal for a commercial and political coalition in general, nor the ensuing negotiations to conclude an agreement ended in success. Now, in conformity with the old idea, the German emperor proposed to the Russian government to discuss the following question: "How can European states which possess oil resources (including Russia) create by their own efforts a special organization, whose goal is to provide European states with liquid fuel and distilled products of crude oil, but without relying upon transoceanic markets?"[34] Without at first even referring to the idea of a monopoly, the Kaiser strove to convince the Russian side that all European countries had an interest in realizing this proposal. Later, however, in a letter to Kokovtsov elucidating the idea of the German proposal, Chancellor T. von Bethmann-Hollweg was forced to concede that the proposed measures "in part concern German legislation," and "in part it is a question of agreements between the most significant private companies or even states" with the goal of "establishing a close mutual tie among the main European producing states (especially Russia, Austro-Hungary, and Rumania) and the German Empire, as one of the main consumers of such heating and lighting products." Bethmann-Hollweg promised Kokovtsov that the petroleum industry of Russia, "especially in the event it expands," will receive "a sure sale on the German market." He also declared that "contracts have already been concluded with the largest Russian industrialists for deliveries." But that was not a very clever lie, for it was naive to think that the tsarist government—with its close ties to Nobel—was not informed about the real situation. At the conclusion of his letter Bethmann-Hollweg expressed the firm hope that "the conclusion of our legislative work" will allow the government to realize the above plan "with all its energy."[35]

A political memorandum by the Russian Naval General-Staff, compiled in early 1912, noted that "of our two main adversaries" (England and Germany), "the most dangerous at the present time is Germany."[36] The German policy provoked growing fears on the part of Russia, and this could not remain unknown to the leaders of the German Kaiserreich. Under these circumstances, it would be naive to expect that the tsarist government came halfway to meet the proposals of Wilhelm II. Indeed, in replying to Bethmann-Hollweg, Kokovtsov did not show the slightest enthusiasm for the Kaiser's proposals.

After long delays and deferrals, on December 7, 1912, the German government finally sent the Reichstag its draft proposal for a kerosene monopoly. In introducing the draft law, the state-secretary for finances (Kühn) laced his speech with hostile remarks about Standard Oil, but he nevertheless invited the American company to participate in future oil deliveries to Germany. Just a few days previously, Teagle sent Gwinner a categorical refusal on any negotiations on this question. Even so, the German side now invited the Americans to join in the negotiations.

How is one to understand this? The answer is to be found in the speech of the state secretary himself. "Have we concluded contracts for deliveries?" he asked. And then he replied that the resolution of this question is only after "the project shows that it is viable." Kühn's statement was equivalent to a concession of the failures that preceded submission of the draft law to the Reichstag, while his statement to the American sounded like a call for assistance. His speech revealed another weak side of the German project: Kühn was forced to concede the presence of profound disagreements in Germany itself among the participants of the future oil company.[37] The growing differences between the Deutsche Bank and the Disconto Gesellschaft made an agreement between them very unlikely.

Apart from Standard Oil, Germany could receive oil products only from the Anglo-Dutch firm of Royal Dutch Shell or the petroleum companies in Russia. But the latter pair of possibilities were virtually nonexistent: negotiations with Deterding and Nobel did not give results, because the development of events removed Germany further and further both from England and Holland, and from Russia. The approach of war could be felt with every passing day, and neither political nor military-strategic considerations could permit Germany to link the fate of its oil deliveries with any of these countries. Is that not why Disconto Gesellschaft suddenly decided to get rid of the Nobel stocks? As a result, by May 1914 its portfolio held only 696,000 rubles out of the original 6 million rubles.[38] Simultaneously, the Deutsche Bank hastened to sell off the 1.5 million rubles of stock in the Russian company "Neft'" that it had bought earlier through banks in St. Petersburg.[39] The approach of war acquired growing importance. This became especially clear after the legislative measures taken in 1912-1913 to strengthen the army and navy as well as the huge appropriations for this purpose that followed.

In the assessment of U.S. naval intelligence, the German fleet needed 60 to 70 thousand tons of oil fuel.[40] But Germany had not succeeded in securing regular oil deliveries and sundering her dependence upon Rockefeller. Therefore, in the final analysis, everything stood in 1914 just where it had been on the eve of the agitation for a petroleum monopoly. The country was on the verge of war, but had failed to solve its deficiency in oil—and at the very time that her future adversaries, the Entente nations, had had considerable success in this regard.

NOTES AND REFERENCES

1. *Moniteur,* January 2/15, 1902; Gerretson, *3,* p. 80; F. Hasse, *Die Erdöl-Interessen der Deutschen Bank und der Direktion der Disconto-Gesellschaft in Rumänien* (Berlin, 1922), p. 42.
2. I. von Baumgart and H. Benneckenstein, "Der Kamp des deutschen Finanzkapitals in den Jahren 1897 bis 1914 für ein Reichspetroleummonpol," *Jahrbuch für Wirtschaftsgeschichte* (Berlin, 1980) *2,* p. 102.
3. Report of the deputy consul-general in Frankfurt-am-Main, J. W. Hanauer (September 10, 1903). In *U.S. Consular Reports, 73,* p. 564.
4. O. V. Brackel and J. Leis, *Der dreißigjährige Petroleumkrieg* (Berlin, 1903).
5. *Neftianoe delo,* 28.VI/11.VII.1904.

6. From the report of the United States consul in Berlin, F. H. Mason (October 2, 1901). In *U.S. Commercial Relations, 1900, 2,* p. 261.

7. A. Vagts, *Deutschand und die Vereinigten Staaten in der Weltpolitik* (New York, London, 1935), *1,* pp. 354-355, 357.

8. U. Brack, *Deutsche Erdölpolitik vor 1914* (Hamburg, 1977), p. 169, 178.

9. See Tables 5 and 6 in the Appendix.

10. Report of the resident "Z" (January 27, 1904 and October 29, 1906) in the United States National Archives, U.S. Office of Naval Intelligence, RG 38, Box 838, File 1623.

11. Ibid.; *Moniteur,* May 5/18, 1904 and January 14/27, 1907.

12. *MNKPR,* p. 364; von Baumgart and Benneckenstein, p. 109.

13. Brack, p. 506; von Baumgart and Benneckenstein, p. 102.

14. Deutsche Bank to E. Nobel (February 28, 1908). In *LGIA,* f. 1258, op. 2, d. .228, ll. 161-162.

15. *Moniteur,* January 1/14, 1907 and February 10/23, 1907.

16. Report from P. Miller (dated February 6/19, 1907). In *AVPR,* f. Agentstvo v Berline, op. 535, d. 51, l. 353.

17. A. Gukasoff to P. Gukasoff (January 10/23, 1905). In *TsGIA AzSSR,* f. 583, op. 1, d. 1259, l. 536.

18. E. K. Grube to Nobel (September 2/15, 1910). In *LGIA,* f. 1258, op. 2, d. 224, l. 81.

19. See D'iakonova, pp. 107-110.

20. V. I. Bovykin, "Rossiiskaia neft' i Rotshil'dy," *Voprosy istorii* (1978), *4,* pp. 30-31.

21. Ibid.

22. T. R. Kin to Nobel (June 28, 1911). In *LGIA,* f. 1258, op. 2, d. 268, l. 134.

23. Reichstag, März 13, 1911 (Bd. 265: 5494).

24. Von Baumgarten and Benneckenstein, p. 90.

25. Lenin, *27,* pp. 368-369.

26. Gibb and Knowlton, 208-219.

27. Brack, pp. 34-35.

28. The agreement between the Nobel Co. and Disconto Gesellschaft (November 22/December 5, 1911). In *TsGIA SSSR,* f. 630, op. 2, d. 297, ll. 7-8.

29. *MKNPR,* p. 531.

30. Report of resident "Z" (August 8, 1908). In the United States National Archives, U.S. Office of Naval Intelligence, RG 38, Box 838, File 1623.

31. *TsGAVMF SSSR,* f. 418, op. 1, d. 661, l. 135.

32. Report from E. A. Behrens (October 24, 1912). In *TsGAVMF SSSR,* d. 2645, l. 59.

33. Von Baumgart and Benneckenstein, p. 119.

34. Kokovtsov to T. von Bethmann-Hollweg (September 16, 1912). In *MKNPR,* p. 538.

35. T. von Bethmann-Hollweg to Kokovtsov (November 25, 1912). In *MKNPR,* pp. 548-549.

36. Spravka Morskogo general'nogo shtaba (February 3, 1912). In *TsGAVMF SSSR,* f. 418, op. 1, d. 3920, l. 23.

37. Reichstag, December 7, 1912, Bd. 286: 2633-2638.

38. "Pamiatnaia zapiska ob inostrannykh kapitalakh v russkoi neftianoi promyshlennosti, prinadlezhaschikh poddannym stran, uchastvuiushchikh v voine protiv Rossii" (1915). In *TsGIA SSSR,* f 37, op. 77, d. 877, l. 1.

39. Record-book of the sale of stocks of the Company Neft' (1913-1917). In *TsGIA SSSR,* f. 1459, op. 1, d. 28.

40. Report of agent "Z" (March 30, 1914). In the U.S. National Archives, U.S. Office of Naval Intelligence, RG 38, Box 838, File 1623.

Chapter V

The Near East and Oil for the British Fleet

THE d'ARCY CONCESSION IN PERSIA

At the very outset of the century, in 1901, the British financier William K. d'Arcy, after accumulating a large fortune from gold-mining in Australia, obtained an oil concession in Persia. Six years later, on the basis of this concession, the Anglo-Persian Oil Company was created, an action that was intimately linked to far-reaching military and strategic calculations. At one level, England sought these concessions in Persia in order to acquire its own oil reserves and thereby compensate for the country's shortage of petroleum sources. But by strengthening its influence in Persia, England also expected to bolster its sway over the strategically important region of the Near East. That is why the British firm in Persia, from the very beginning, found itself under the vigilant control of the Foreign Office.

The history of the negotiations leading to the Persian concession to d'Arcy, conducted in Teheran by his representative A. L. Marriot, has received considerable attention in the secondary literature. In the early 1980s G. Jones and R. Ferrier published two well-researched monographs that drew upon previously unknown archival materials.[1] One cannot, however, accept all of the authors' conclusions. In particular, it is very difficult to understand why Ferrier seeks to refute the account given by the English emissary in Persia, A. Hardinge, who was charged by the Foreign Office with the task of assisting the d'Arcy enterprise.

In his memoirs Hardinge wrote that obtaining an important oil concession for the British company was his "first important duty" in service.[2] The English government feared that the attempt to receive a concession in Persia might provoke opposition from Russia. The British plan did run contrary to the plans of Russian oil entrepreneurs, who had a strong economic interest in the Near Eastern markets. But most important, England's effort to assert influence in Persia elicited dissatisfaction in St. Petersburg because of political considerations. The English therefore conducted the negotiations for concessions in strictest secrecy. Fearing opposition from Russia and seeking to deprive the tsarist government of any formal grounds for protest against this deal between England and the Persian government, the British diplomats devised a clever ruse that Hardinge recounts in his memoirs. Namely, the English decided to inform the Russian emissary Argiropulo in writing of the terms in the planned agreement—but composed the letter in a language unknown to Argiropulo himself. The sole person in the Russian mission who could read it was the translator, Schritter, who at the time was off on a hunting expedition.[3] With Schritter away, the English managed to complete the negotiations and to sign the contract. After this the representative of Premier Amin al-Sultan visited Argiropulo and informed him that the agreement had been made.

Ferrier calls the Hardinge account "an imaginative tale" and asserts that his "memory played him false." He rejects the possibility of such a device on the grounds that the Hardinge version is not confirmed by a single other contemporary. In any case, his work in the archives did not turn up any traces to substantiate the Hardinge account.[4]

To the "imaginative tale" of Hardinge, Ferrier juxtaposes his own version, with the comment that "the truth is more exciting." In his view, Amin al-Sultan was "well known for his pro-Russian sympathies" and did not wish to harm his relations with Argiropulo.[5] He had a vested interest in maintaining these relations, since the negotiations for an English concession were conducted at the very height of talks for a new loan with Russia.[6] To support this version of events Ferrier cites a report from Hardinge to the Foreign Secretary Lansdowne: "Great secrecy was observed in preparing the necessary papers, as it was felt that if the Russian Legation got news of the project, it would attempt to crush it ..." The report shows, however, that information about the negotiations had reached Argiropulo.[7]

Ferrier notes that, insofar as information had been leaked, there was no point in denying the fact that negotiations were being conducted. But the memoirs of Hardinge contain no such assertions. It merely tells how they informed the Russian emissary of the terms of the concession, not about the fact that negotiations were underway.

The English historian supplements the Hardinge report to Lansdowne with an account based on the now-famous memoir of A. Kitabgi-Khan, a director of the customs administration who participated in the negotiations as a mediator. However, the information in that source hardly gives grounds to reject the memoirs of the British diplomat. During the negotiations the Persian premier asked Kitabgi-Khan: "Do you understand what is at stake? They are capable of informing the Russians." After Argiropulo had learned of the negotiations and asked Amin al-Sultan for an explanation, the latter replied that his fears were groundless and further averred that "... the conclusion of this affair, if it ever took place, would take many months, and that during that time, he would have plenty of leisure to inform his government and to demand instructions." During this time the emissary "can, without any haste, inform his government and ask for instructions." This was said on May 24, 1901. But two days later (May 26th) Argiropulo again visited Amin al-Sultan, this time at home, and recommended that he reject the English concession. Nevertheless, the agreement was signed on May 28th.[8]

All these details are fully consistent with the account in Hardinge. Evidently, the document to which the diplomat refers in his memoirs (the terms for the d'Arcy concession) were transmitted by the Russian mission after the meeting between Amin al-Sultan and Argiropulo on May 26th. In any case, such a step will appear completely logical from the perspective of those who brought the contract to the point of signing. The absence of archival documentation to confirm the Hardinge account is not sufficient grounds to dismiss it as an invention.

Indeed, the very character of the negotiations was such that it was all limited to an oral understanding among a narrow circle of people and did not leave traces in the paperwork. Insofar as much was at stake, no information was leaked and, were it not for the Hardinge memoirs, no one would ever have learned of it. At any rate, this interpretation seems more probable than the version offered by Ferrier.

As is well known, the British government secured the concession for d'Arcy with the help of diplomatic pressure and a huge monetary bribe to the Shah and his retinue.

After the Russian representative learned the contents of the agreement for a concession, Argiropulo made a vigorous protest. He expressed his "profound amazement over the unseemliness of the secrecy that has been hitherto observed" and pointed "to the direct mistake of giving rights on such broad principles." Argiropulo then repeated this protest personally to Amin al-Sultan, who (as the Russian emissary reported) "seemed very embarrassed and did not justify himself."[9] The point is that, from the late 1890s, the Shah's government made wide use of Russian financial support, thereby freeing itself from the influence of England's Imperial Bank of Persia that had previously enjoyed a monopolistic position. Under these conditions, the agreement with d'Arcy appeared to be disloyal toward Russia.

When news of the d'Arcy concession reached St. Petersburg, it provoked a storm of indignation. The Ministry of Foreign Affairs and especially Witte's Ministry of Finance perceived this act not only as an encroachment on Russia's political influence in Persia, but also as a blow to her economic interests. Fears were also raised for the fate of the Baku oil trade, which by this time held a virtual monopoly in the Persian Gulf, comprising approximately 90 percent of oil shipments.

Although point 7 of the agreement specifically excluded the northern provinces of Persia (where Russian influence was predominant) from the British sphere of activity, the same point also guaranteed the British monopoly rights for laying pipelines through the southern part of the country.[11] However, as early as the 1880s, there had been discussions in Russia about the construction of a pipeline to a port on the Persian Gulf, the goal being to cut the cost of transporting oil from Baku. That project served not only the needs of Persia but other countries of the East, if they were to avoid the significantly higher costs of shipping the oil through the Suez Canal.

To be sure, in 1884, when a proposal for a Trans-Persian pipeline was first discussed, a Special Conference decided to take no action on the scheme. But in the summer of 1901, amidst the crisis in Russia's oil industry, the project was again revived because of the need to expand "the market for selling our kerosene in Asia, where we can easily hold a brilliant position" and where, in the event a pipeline was built in order to reduce shipping costs, it will be impossible for rivals of Russian oil to compete. The new project was supported by the masters of Russian oil—Nobel and the Rothschilds.[12]

The tsarist government subjected Persia to intense diplomatic pressure in an effort to extract a concession to construct the pipeline. Especially insistent in pursuing this concession was the Ministry of Finance and Witte personally, who demanded that the concession of a pipeline be made a precondition for offering Persia new Russian loans.

Although tsarist Russia itself chronically experienced an acute need for monetary means and had to seek loans on international markets, in Persia she played a quite different role. From the late 1890s, the Russian Discount and Loan Bank of Persia became the chief creditor for the Persian government. And the loans were an important instrument for the tsarist government to exercise economic and political influence on the East.[13]

In September 1901 the Russian government approved the issue of the next loan to Persia. But Witte made one of its conditions the grant of the pipeline concession.[14] In addition, the minister of finance directed his agent in Teheran (E. K. Grube), if necessary, to offer Amin al-Sultan a monetary bribe—equal to that which he had received (as Petersburg had heard) for making the concession deal with d'Arcy.[15]

However, all these attempts proved unsuccessful. In November 1901 Argiropulo sent a message from Teheran, that the English were prepared "to resort to extreme measures for the sake of securing the rights they have reserved."[16] When Russia began threatening to withhold the loan, the English representatives offered credits themselves. This proposal was first advanced by the British emissary Hardinge to the Persian premier, and then to the Shah himself through his English doctor, Adcock.[17]

Sometime later the British ambassador in St. Petersburg, Charles Scott, gave the Russian foreign minister, Count V. N. Lamzdorf, a note demanding that the Russians abandon their claims to an oil pipeline. He declared that "to avoid any possible complications," he was "empowered" by Marquise Lansdown, the head of the Foreign Office at the time, to "remind" Count Lamzdorf that the construction of any pipelines in the projected direction "was already granted by the Persian government to the British subject W. N. d'Arcy in May, 1901."[18]

Nevertheless, after receiving the memorandum from Lamzdorf, Witte asked that it be pointed out "to Sir Charles Scott that the concessions, which he has in view, were solicited by a private institution, the Discount and Loan Bank of Persia, and therefore the Russian government is not in a position to give any kind of precisely explanations as to its content."[19] This formal reply was highly disingenuous—not only because the government actively participated in the negotiations for the pipeline, but also because that bank was actually a branch of the State Bank of Russia (similar in status to the Russo-Chinese Bank). The sole difference was the capital of the former consisted almost entirely of state contributions, while that of the later included the participation of Russian and even foreign capital. None of this, of course, was a secret to the English side.

Ignoring the text of the English concession, the Russian representatives insisted that the project of the Discount and Loan Bank pertains "exclusively to the export and transport of local oil."[20] However, the Foreign Office instructed Hardinge not to compromise. Although the instructions sent from London noted that the British government will not obstruct Russian commercial access to the Persian Gulf, it did not extend this decision to include a pipeline.[21] A firm warning was issued to the Teheran government, and the situation of the Persian premier became so complicated that he was threatened with removal and disfavor.[22] This outcome was undesirable for Russia, since a replacement of the premier would play right into the hands of the English. And, added to all that, in February 1902 d'Arcy himself made an offer through the Shah's retinue to make a loan for 200,000 £ St., and the latter, in view of the need for money during his impending trip to Europe, was inclined to accept. In telegraphing St. Petersburg about this, Grube reported that the English loan was offered to the Shah "in payment for the oil concession." He did not assess the chances of a Russian proposal very optimistically, declaring that it was impossible "to attain the pipeline concession from the Shah at the present time," and complained of the "unceasing intrigues of the English emissary in Teheran against the conclusion of a Russian loan."[23] Argiropulo sent Lamzdorf a telegraph with essentially the message. "The combination that d'Arcy has offered the Shah (consisting in the issue of 200,000 £ St.)," he wrote, "has significantly altered the situation, but not in our favor. The Shah has fallen completely under the influence of his retinue and thinks only about his journey." The envoy advised his government to abandon demand for a pipeline concession and to grant the law without any conditions.[24] Under these circumstances, further aggravated by the deterioration of international relations in the Far East, Russia could not risk endangering its ties with England and was forced to abandon the project for a trans-Persian pipeline and to offer the loan to Persia without any conditions.

Having decided to avoid straining its relations with England, Russia also encountered a rising danger from Germany, the significance of which had already been assessed in both London and St. Petersburg. The interests of both powers were threatened by Germany's intrusion into Turkey and, in particular, the construction of the Baghdad Railway, which set for itself the goal of asserting German hegemony in the Near and Middle East. In this connection, in mid-1902 Witte issued a special memorandum, which stressed the need for decisive measures "to paralyze the significance of the Baghdad Railway, thanks to which Germany threatens to become a dangerous rival in the East both for us and for the English."[25] All this, taken together, had the effect of putting an end to the conflict between England and Russia over the question of the d'Arcy concession.

Nevertheless, English circles continued to take a keen interest in the fate of the oil enterprise in Persia. Not only the Foreign Office but also the British Admiralty maintained a vigilant supervision over the activities of the d'Arcy company. It was precisely at the turn of the century that the Admiralty began to discuss the question of upgrading the navy's fuel—that is, re-equipping the naval forces with oil-burning engines to replace those using coal.[26]

By the end of 1903 the general sum of expenditures by the English company on explorations for Persian oil amounted to 160,000 £ Sterling, but no tangible results were achieved. In July 1903, in a letter to his aid Jenkin, d'Arcy noted that he would like to have some idea of the prospects for further oil exploration, since "every purse has its limits," and he added that he knows well "the limits of my own."[27] At that moment d'Arcy was in Karlsbad, where he met Admiral J. Fisher, who had come there to relax, and who never tired of repeating: "Whoever controls oil will rule the world."

Fisher and d'Arcy easily found a common language. In addition, the admiral showed such a great interest in d'Arcy's enterprise that he joked, should he not go to Persia instead of returning home (the royal naval base in Portsmouth) after his holiday? Although the Admiralty's interest in providing oil to England had been evident earlier, the meeting at Karlsbad impelled d'Arcy to apply—with the Admiralty's support—to the British government for financial assistance. At the end of 1903 he filed the appropriate request, requesting the sum of 120,000 £ Sterling. But the government turned down the request. D'Arcy next convoked a meeting of the company's board of directors and warned that he had exhausted all his resources and was forced to appeal to an outside party for financial assistance. After an unsuccessful attempt to obtain a loan from the English bank of Joseph Lyons and Co., d'Arcy began negotiations with the Parisian Rothschilds. He met with Baron Alphonse and Jules Aron, and in February 1904 held further negotiations with them on the French Riviera. But here too his efforts ended in failure. He then applied once more to British financiers, this time to E. Cassell, a friend of King Edward VII—but again in vain. He even explored the possibility of obtaining a loan from the Americans. But he was given to understand that they "would have nothing to do with the matter till oil is found."[28]

Then a specially created commission at the British Admiralty, headed by the civilian Lord J. Pretimen, which had explored the question of a possible expansion of British oil holdings, joined in the search for a creditor. It recommended that the company Burma Oil, a contractor of the British government, to offer financial assistance to the d'Arcy enterprise. In May 1905, with the participation of Burma Oil, a Concession Syndicate was formed and allocated means to continue the oil exploration in Persia.[29]

Then d'Arcy encountered as well another kind of difficulty, which the British government once more helped to resolve. It was necessary to establish order among the local population, on whose territory the explorations were being conducted. For this purpose it proved necessary to make an agreement with the Bakhtiari khans, who were allocated a 3 percent share of all the oil enterprises and a special payment for protecting the company's property. The agreement was signed in November 1905 with the assistance of the English consul in Arabistan.[30]

Thanks to these measures, d'Arcy continued his exploratory work and finally, on May 26, 1908, a fountain of oil erupted from an oil well in Musjid-i-Sulaiman, marking the discovery of one of the largest oil reserves in the world. On the eve of this event the English enterprise in Persia had again experienced serious financial difficulties. The situation seemed so hopeless that on May 14, 1908 an order was sent to the oil prospecting team to cease drilling if oil was not discovered by a depth of 500 meters.[31] A few days later, from a depth of 378 meters, a powerful geyser of oil burst skywards. Nevertheless, this event did not evoke a boom on the stock exchange.

FORMATION OF THE ANGLO-PERSIAN OIL COMPANY

In the second half of 1908 intensive negotiations were initiated to reorganize the d'Arcy company and its system of financing, which culminated on April 14, 1909 in the formation of the Anglo-Persian Oil Company, with a capital of 2 million £ Sterling. Of this capital 1 million £ Sterling in common stocks were shared by Burma Oil (570,000 £), the Concession Syndicate (400,000 £) and Lord Strathcona (30,000 £), who became chairman of the board in the new company. Another 1 million £ in preferred stock were intended for sale.[32]

D'Arcy participated in the creation of the new company, but lost his former influence. In compensation for his expenditures he received 170,000 shares of stock from Burma Oil with a value of approximately 900,000 £ sterling, and also the an honorary position as directory of the company. In a private letter his wife wrote that she was "disappointed to see they had left my husband's name out of it absolutely."[33] The British petroleum enterprise in Persia, in any event, was transferred to more reliable hands. In terms of its financial capacities and political connections, Burma Oil was doubtless more influential than the d'Arcy company.

CONVERSION OF THE BRITISH FLEET TO PETROLEUM FUEL

Against a background of deteriorating international relations and an accelerating arms race in the years preceding World War I, England showed a growing interest in strengthening its strategic position in the Near East.

By 1912 the petroleum storage tanks of the British fleet contained approximately 200,000 tons. That was not a small amount, but in the event of war and conversion to liquid fuel, it did not represent a significant quantity: according to preliminary calculations, the annual consumption of oil under war conditions would constitute about 1 million tons. The Russian naval agent in London, L. B. Kerber, noted that "the absence of liquid fuel in the heart of England itself and the fear that in wartime oil shortages will appear," almost forced the government to preserve the system of coal heating.[34] However, the naval arms race demanded a reconstruction of naval vessels. According to the assessment of the U.S. Secret Service, the British Admiralty was already "convinced" that oil is "the ideal fuel."[35] Therefore the conversion of the fleet to petroleum fuel became a key imperative.

When Winston Churchill was appointed first lord of the Admiralty at the end of 1911, he put this issue at the top of the naval agenda. Use of petroleum fuel would permit him to construct faster vessels and to arm them with heavier weapons. Churchill called the advantage from applying petroleum fuel "inestimable." "In equal ships," he noted, "oil gave a large excess of speed over coal." In addition, it would expand the radius of operation for naval ships by 40 percent. Ships using oil could replace fuel at sea by refilling from tankers. They could remain at their bases, free from the necessity of constantly sending a quarter of the ships to port for more coal. "The use of oil," noted Churchill, "made it possible in every type of vessel to have more gun-power and more speed for less size and less cost."[36]

The question of converting the fleet to oil was debated in the context of mounting Anglo-German antagonism. Germany's program for ship construction accelerated the naval arms race. England declared that, for the construction of each German battleship, it would react by building two new vessels. The British naval program foresaw the construction of super-powerful vessels, whose fighting capabilities would be improved by conversion to oil.[37] Many English ships of a lighter type—destroyers and submarines—had already been reequipped to use oil. In 1912 the general number of active or soon-to-be launched vessels of this type numbered 150. In another 18 to 24 months this number was to double.[38] In April 1912 Churchill raised the question of converting the naval forces to liquid fuel. Although he did not immediately succeed in winning the approval of the British cabinet, he eventually managed to persuade the majority of his view after enlisting the support of the king, Edward VII.

Although even some members of the Admiralty had reservations about Churchill's views, here too he eventually succeeded in winning the upper hand. He was decisively supported by Admiral Fisher, who in 1910 left the very post that Churchill now held. Immediately after Churchill's appointment, the "oil maniac" came to the new head of the Admiralty and insistently recommended that he resolve the question of converting the naval forces to oil. In the contrary event, he warned, "a German motor battleship" will humiliate "our 'tortoises'!"[39] In June 1912 Churchill sent a letter to the retired admiral with the request that he head a special royal commission: "You have got to find the oil; to show how it can be stored cheaply; how it can be purchased regularly and cheaply in peace; and with absolute certainty in war. Then by all means develop its application in the best possible way to existing and prospective ships." Churchill gave Fisher the possibility, as he put it, "to crack the nut," although he warned the admiral that "Your Royal Commission will be advisory and not executive. It will assemble facts and state conclusions. It can not touch policy or action."[40]

Insofar as the Anglo-Persian Oil Company had still not begun commercial activity, Fisher proposed to make Deterding the first preference in supply petroleum. In a letter to Churchill he called Deterding "Napoleon and Cromwell all rolled into one." He avowed that "he is the greatest man I have ever met" and warned: "Placate him, don't threaten him." Churchill, however, took a somewhat different position.[41]

Deterding really did have great power and expressed a readiness to collaborate with the Admiralty, but he was not willing to limit himself to being supplier for the royal fleet alone. In August 1912 he gave this warning to his English partners: "We must not push our chauvinism too far," meaning that the sphere of interests for Royal Dutch Shell should be the entire world. Just at that time Deterding bought up the Rothschilds' oil enterprises in Russia and was fighting Rockefeller for control of the oil fields of the United States. Appearing before Fisher's commission in early 1913, he declared: "I always want the utmost profit possible." Deterding advised the Admiralty to build reservoirs in various regions of the world, which his company could then pump full of oil at previously set terms.[42]

Deterding's British partner, M. Samuel, declared that Royal Dutch Shell is ready to deliver oil to the English fleet alone, but on "the unconditional term" of "Foreign Office and diplomatic support for [Shell] oil interests" in various parts of the world. But Deterding did not wish to be bound by such obligations, not wanting that his firm be anglicized, and therefore disavowed Samuel's statement. "I am entirely internationalist," he wrote later to one of his contractors.[43] Royal Dutch Shell maintained relations with the Deutsche Bank, and this served as a source of constant reprimands from the Admiralty.

Subsequently, in speaking of the government's approach to the problem of supplying the fleet with oil, Churchill noted that the government's policy was "twofold," having both "interim" and "ultimate" lines. The "interim"" line foresaw the making of contracts to "secure a regular and an adequate supply during this immediate future period at reasonable and steady prices." This category included contracts with Royal Dutch Shell. The "ultimate line" aimed to ensure that "the Admiralty should become an independent owner and producer of its own supplies of liquid fuel." Churchill declared that "we must become the owners, or at any rate, the controllers at the source of at least a proportion of the supply of natural oil which we require."[44] The "ultimate line" of the Admiralty amounted to solving this goal with the assistance of the Anglo-Persian Oil Co.

On February 18, 1914 the Cabinet of Ministers made a decision to purchase the Anglo-Persian company. On May 20th the Admiralty concluded an agreement with the company, which increased its capital from 2 to 4 million £ Sterling. The government acquired the controling share of stock for 2.2 million £ St. This sum was invested in exchange for a 30-year contract for the regular supply of Persian oil to the British float at moderate prices. The ruling board added two representatives from the government, who were given the right of "veto" in the resolution of all strategic questions.

During the debates in parliament, a number of deputies (with Samuel as their leader) spoke against ratification of the agreement with the Anglo-Persian Oil Company because of the indefensibility of its holdings, given the international competition in the Near East. It would be more correct, in their view, to rely upon an oil company operating within the realm of the British Empire, having in mind, of course, Royal Dutch Shell. Samuel reproved Churchill for ignoring the other British company: why, for example, did they choose Persia in lieu of Egypt, where by this time Royal Dutch Shell had already established a branch firm?[45]

The government easily beat back these attacks by assuring parliament that no one was threatening the British positions in Persia. The agreement with the Anglo-Persian company was the sole possibility for breaking out of the circle of oil trusts, which strove to use a monopoly with the aim of sustaining high prices. In the vote in the House of Commons, 254 deputies supported ratification of the agreement, with only 18 voting nay.[46]

When one of the directors of the Anglo-Persian Company asked Churchill the following day, "how did you manage to carry the House with you so successfully," the latter replied that it was "the attack on monopolies and trusts that did it."[47] The British press portrayed Churchill virtually as a "trust-buster," although in fact there were no grounds for that claim. While using the methods of anti-trust demagoguery, Churchill at the same time did some curtsies in the direction of Royal Dutch Shell. He declared that he did not seek to discredit the company, with which the government had never "quarrelled," to which Samuel retorted: "Of course." Churchill conceded that "we have always found" the representatives of Shell "courteous, considerate, ready to oblige, anxious to serve the Admiralty and to promote the interests of the British Navy and the British Empire—at a price. The only difficulty has been the price."[48]

"We lead a stormy life in our interviews with the 'Shell,'" noted F. Hopwood (one of the high-ranking figures in the Admiralty), "but I always try to keep on good terms with them myself, because I feel we want their stuff."[49] There is no doubt that Churchill did not wish to sunder these ties. It was no accident that, during his attacks in Parliament, he personally excluded Deterding. After the debates had ended, through his brother, Churchill sent Deterding a stenographic copy of the debates (with a postscript), from which one can conclude that he deliberately avoided mentioning his (Deterding's) name. More importantly, the Admiralty left in force its earlier contracts with Royal Dutch Shell and regularly renewed them when their term had expired. In addition, Churchill gave orders to obtain contractual relations with Deterding on a long-term basis.[50]

Under the accompaniment of anti-trust declarations, the British government signed the treaty with the Anglo-Persian Oil Co., thereby opening the way to the later creation of the oil monopoly British Petroleum—presently one of the six largest multinational corporations for the production, distillation and trade of oil in the capitalist world. In the early 20th century, under conditions of explosive growth in monopolistic conglomerates, direct participation by the state in a large commercial-industrial enterprise appeared rather extraordinary. It was an unprecedented case, although England had its reasons for this decision, not all of which were related to the problem of supplying oil for its naval forces.

The military preparations and mounting conflict between the Entente countries and the Austro-German bloc stimulated a maturation of modern forms of the oil business. It should be emphasized that the military factor exerted a great influence on the structure and character of industrial corporations. On the eve of World War I, England lagged behind such "advanced" countries of modern capitalism as the United States and Germany, where the predominance of trusts had already become established. For a long time, England compensated for the more "outmoded" forms of capitalist organization by relying upon her colonial monopoly. However, because of the increasing internationalization of the world economy, British corporations began to experience certain difficulties. They proved weaker than their rivals, and the English government sought to strengthen them through various measures of state support. It is in this sense that one should consider the financing of the Anglo-Persian Co., which now had to encounter the most powerful commercial-industrial conglomerates in the world—the oil trusts. Financial assistance to the company was the highest manifestation of state support under the conditions of the impending world war. This ostensibly economic measure bore a clear political coloring.

The political dimension of the British oil enterprise in Persia was obvious from the moment that d'Arcy obtained the concession and became still more evident in the agreement with Anglo-Persian Oil Co. in 1914. The British government used the company as an instrument for its policy in the Near East and, indeed, not only in Persia. An example for that is offered by negotiations to establish a Turkish Petroleum Co., the concluding phase of which would coincide with the government's decision to finance the Anglo-Persian company. At the insistence of the Foreign Office, the Anglo-Persian company was allotted the controlling bloc of shares in the Turkish Petroleum Company.[51]

The initiative for the negotiations toward a concession in Turkey belonged to the Deutsche Bank, which thought that its right to oil exploitation had been secured through the contract to construct the Baghdad Railway line. In April 1909 a group of influential English financiers, headed by E. Cassel, established the National Bank of Turkey. At the advice of the English government, H. Babbington Smith (England's representative in the Administration of the Ottoman debt) became president of the new bank. He was acquainted with the head of the Deutsche Bank, A. Gwinner, and through him with the German Chancellor, Theobald von Bethmann-Hollweg. Negotiations were begun, and after difficult bargaining under the mediation of the oil broker Calouste Gulbenkian, in October 1912 they came to an agreement on the creation of a Turkish Petroleum Co. The company's capital in stocks amounted to 80,000 £ Sterling, of which 25 percent belonged to the Deutsche Bank, 35 percent to the National Bank of Turkey (or the English financiers represented in it), and the remaining 40 percent remained "at the discretion" of Gulbenkian, who offered 25 percent to Deterding and kept 15 percent for himself. Babbington Smith became the company's president.[52]

At the end of 1913, the British government proposed to the National Bank of Turkey that it transfer its share to the Anglo-Persian Oil Co. An analogous proposal was sent to Deterding, but it only succeeded in making him "furious." Deterding declared that "he will never agree with this." In early 1914 a conference of interested parties was convoked, and on March 16th they signed a new agreement, according to which the company's capital was doubled, with 50 percent of the stock being allotted to the Anglo-Persian Oil Co. Accepting Fisher's council not to quarrel with Deterding, Churchill ordered that the 25 percent assigned to Royal Dutch Shell remain unchanged. Deterding appreciated this gesture and sent Churchill a letter of thanks: "Whenever I can reciprocate by being of any service to you, I shall feel most happy to render this unhesitatingly." The Admiralty continued to conclude contracts with Royal Dutch Shell.[53]

Under these circumstances, the English government decided that the participation of the Deutsche Bank did not represent a threat, and therefore did not object to leaving 25 percent of the stock for the Germans. Gulbenkian, whose inventiveness was responsible for the creation of the Turkish Petroleum Co., was simply dropped from the game at this point. He was not admitted to the negotiations and, as compensation for his earlier "services," the Anglo-Persian Company and Royal Dutch Shell allocated him 5 percent of their own stocks—from which comes Gulbenkian's nickname of "Mr. Five Percent." The result of all these negotiations was that the largest bloc of stocks and the deciding word in the affairs of the Turkish Petroleum Co. now belonged to the Anglo-Persian Oil Company. And this is precisely what the British government had sought.

Soon after the agreement on the Turkish Petroleum Co. had been signed, the Turkish government officially promised to grant the Germans a concession. The significance of that promise, however, was virtually nullified in April 1914 by the agreement between the Admiralty and the Anglo-Persian Oil Company. Cries were to be heard in the Reichstag and the press demanding that England not be given advantages in the naval arms race. During the next cruise on his yacht, Kaiser Wilhelm II (in the presence of Admiral Tirpitz) subjected the Deutsche Bank director, Gwinner, to withering criticism: "There he is now, the man who supplies our precious fuel to foreign navies!"[54]

The actions of the British government toward the Anglo-Persian Oil Co. also failed to evoke jubilation among England's allies in the Entente. They also expressed dissatisfaction with the fact that England had used the company as a weapon for their own policy. The French ambassador to London, P. Cambon, informing his government about the actions of the Admiralty, expressed the fear that "in the future a significant part of the oil production may prove to be in English hands."[55] In Russia, the newspaper *Novoe vremia* sharply criticized the contract with the Anglo-Persian Oil Co., calling this contradictory to the letter and spirit of the Anglo-Russian Agreement of 1907.[56] This statement was probably inspired by the Ministry of Foreign Affairs and provided grounds for formal note to the English government demanding "an explanation."[57] Tsarist diplomats wanted England to renounce "her rights to exploit oil in the part of the neutral zone, contiguous to our sphere of influence, and especially in the latter."[58] The Russian ambassador in London, A. K. Benkendorf, was instructed by the minister of foreign affairs to meet with the head of the Foreign Office (E. Grey), and during their discussions the latter noted that the agreement does not introduce anything new and that "no kind of changes" had transpired.[59] He said the same thing in Parliament, and afterwards informed Benkendorf that he "understands the need to give an explanation to the Russian press."[60] After this Grey met with the correspondent of the paper *Peterburgskii kur'er,* B. Lebedev, and presented his point of view.[61] In addition, he directed the British ambassador in London, G. Buchanan, to transmit to the Russian government a note, which declared: "The government of His Majesty has not acquired any rights to concessions which the Anglo-Persian Oil Co. did not possess at the time when the Anglo-Russian convention was made."[62] However, all these reassurances failed to change the dim view that high circles in St. Petersburg took of the new British actions in Persia.

Following Parliament's ratification of the agreement with the Anglo-Persian Oil Company on June 17, 1914, the contract acquired the force of law. The requisite financial appropriations foreseen in the contract were approved on August 5, 1914, a day after the onset of World War I. That was profoundly symbolic, because the preparations for war, as we have seen, was one of the reasons for the purchase of the company, the foundation of which was the original concession to d'Arcy.

NOTES AND REFERENCES

1. G. Jones, *The State and the Emergence of the British Oil Industry* (London, 1981); R. Ferrier, *The History of the British Petroleum Co.* (London, 1982).

2. A. Hardinge, *A Diplomatist in the East* (London, 1928), p. 278.
3. Ibid., p. 279.
4. Ferrier, p. 95.
5. Ibid.
6. B. V. Anan'ich, *Rossiiskoe samoderzhavie i vyvoz kapital0v 1895-1914 gg.* (Leningrad, 1975), p. 36.
7. Ferrier, pp. 39-40.
8. Ibid., p. 40.
9. Argiropulo to V. N. Lamzdorf (June 29/June 12, 1901). In *TsGIA SSSR,* f. 560, op. 28, d. 247, l. 7.
10. B. V. Anan'ich, "Rossiia i kontsessiia d'Arsy," *Istoricheskie zapiski,* (Moscow, 1960) *66,* p. 287-289.
11. Ibid.
12. Ibid.; H.J. Wigham, *The Persian Problem* (London, 1903), p. 267.
13. B. C. Anan'ich, "Uchetno-ssudnyi bank Persii v 1894-1907 gg.," *Monopolii i inostrannyi kapital v Rossii* (Moscow-Leningrad, 1962), pp. 277f.
14. Anan'ich, "Rossiia i kontsessiia d'Arsy," pp. 284f.
15. Witte to E. K. Grube (February 12/25, 1902). In *TsGIA SSSR,* f. 560, op. 28, d. 247, l. 115.
16. Argiropulo to Lamzdorf (November 15/28, 1901). In *TsGIA SSSR, 1,* p. 49.
17. Telegram from Grube (November 2/15, 1901). In ibid., l. 18.
18. Memorandum from the ambassador of Great Britain in St. Petersburg (January 31/February 31, 19020. In ibid., l. 78.
19. Witte to V. N. Lamzdorf (February 4/16, 1902). In ibid., l. 79.
20. Ibid. This argumentation was prepared still earlier—in the instructions from the head of the State Bank, E. D. Pleske to the director of the Uchetno-ssudnyi Bank Persi, E.K. Grube. See the telegram from Pleske November 5/18, 1901, (ibid., l. 22).
21. Lansdown to A. Hardinge (January 6, 1902). In *British Documents,* 4, pp. 369-372.
22. Argiropulo to Lamzdorf (December 15/28, 1901, February 20/March 5, 1902). In *TsGIA SSSR,* f. 560, op. 28, d. 247, ll. 49, 141.
23. Grube to Witte (February 20/March 5, 1902). In ibid., l. 136.
24. Argiropulo to V. N. Lamzdorf (February 20/March 5, 1902). In ibid., l. 141.
25. See Witte's memorandum (August 3/16, 1902). In *AVPR,* f. Politarkhiv, d. 1002, ll. 502-506.
26. See Jones, p. ix.
27. Ferrier, p. 59.
28. Ibid., pp. 60-62; H. Longhurst, *Adventure in Oil. The Story of British Petroleum* (London, 1959), p. 23.
29. Ferrier, pp. 69-73.
30. B. Schwadran, *the Middle East, Oil and the Great Powers* (New York, 19550, p. 18; ;Ferrier, pp. 75-78.
31. Ferrier, p. 88.
32. Ibid.; Jones, pp. 138-139.

33. Ferrier, p. 98.
34. Report from the Russian naval attaché in England, L. B. Kerber (May 20/June 3, 1909). In *TsGAVMF SSSR,* f. 418, op. 1, d. 3277, l. 38.
35. Memorandum of agent X (January 15, 1908) in U.S. National Archives, U.S. Office of Naval Intelligence, RG 38, Box 837, File 1908.
36. W. Churchill, *The World Crisis, 1911-1914* (London, 1923), p. 129.
37. A. J. Marder, *From the Drednaught to Scapa Flow. The Royal Navy in the Fisher Era, 1904-1909* (London, 1961), *1,* p. 269-270.
38. *Parliamentary Debates,* July 17, 1913, vol. 55: 1466; report of the Russian naval agent attaché in England, Rein (December 2/15, 1912). In *TsGAVMF SSSR,* f. 481, op. 1, d. 2645, l. 24.
39. R. S. Churchill, *Winston S. Churchill,* vol. 2, pt. 3 (London, 1969: 1927.
40. Ibid., p. 590.
41. Ibid., p. 1948.
42. Jones, p. 45.
43. Ibid., p. 45, 73.
44. *Parliamentary Debates,* July 17, 1913, *55,* 1474-1475.
45. Ibid., June 17, 1914, *63,* p. 1137, 1222.
46. Ibid., pp. 1151-1153, 1178-1189, 1250.
47. Ferrier, p. 199.
48. *Parliamentary Debates,* June 17, 1914, *63,* p. 1150.
49. Jones, pp. 171-172.
50. Gerretson, *4,* p. 293.
51. M. Kent, *Oil and Empire. British Policy and Mesopotamian Oil, 1900-1920* (London, 1976), pp. 74-89.
52. Hewins, p. 75.
53. Ibid., pp. 81, 83-84; Jones, p. 172.
54. Gerretson, 4, 283.
55. P. Cambon to G. Dumerque (May 27, 1914). In *Documents diplomatiques française* (Paris, 1936), ser. 3, 10, p. 430.
56. *Novoe vremıa,* July 6/19, and July 7/20, 1914.
57. Rutkovskii to Bark (June 27/July 6, 1914). In *Mezhdunarodnye otnosheniia v epokhu imperializma,* Ser. III, IV, pp. 142-144.
58. Sazonov to Buchanan (June 27/July 10, 1914). In ibid., p. 203.
59. Benkendorf to Sazonov (June 12/25, 1914). In *AVPR,* Persidskii stol, d. 4169, l. 205.
60. Benkendorf to Sazonov (June 17/30, June 18/July 1, 1914). In ibid., ll. 211-212.
61. *Peterburgskii kur'ier,* July 8, 1914.
62. Memorandum posol'stva Velikobritanii v Peterburge (June 21/June 4, 1914). In *AVPR,* Persidskii stol, d. 4169, l. 214.

Chapter VI

The Mexican Drama

One of the most dramatic situations on the eve of World War I was unfolding in Mexico. Against a background of mounting political rivalries in the first decade of the century, and then the Mexican Revolution of 1910-1917, the great powers were engaged in a bitter struggle to gain control of Mexico's oil resources. The competition of American, English, and German corporations for Mexican oil fields and railway lines was intimately intertwined with the struggle against the Mexican Revolution—the first great national-liberation movement of the twentieth century in the Western hemisphere.

Oil production on an industrial scale began in Mexico in 1900. Among the more active claimants to the exploitation of oil resources was the Rockefeller firm of Waters-Pierce Oil Co. In addition, the American Edward Doheny established the Mexican Petroleum Co. of California, which owned enormous tracts of land, oil fields, pipelines, storage facilities and a large fleet of tank cars. Doheny maintained close ties with Rockefeller. In 1912 the Senate Commission on Foreign Affairs discussed the Mexican problem. According to the general opinion, declared one of the participants of this discussion, Rockefeller controls the Doheny company, although it does not belong to him. The commission learned that, when Doheny and his partners found themselves in financial difficulties, they went to Broadway 26 and received 400,000 dollars.

American companies made it their goal to achieve control over Mexico's oil reserves. And this plan was conceived as a large-scale operation, including the establishment of control over the country's railway network. By exploiting the financial difficulties of Mexico, American capital attempted in 1906-1907 to take the railroads into its own hands. However, in 1907-1908 the Mexican government reorganized the administration of railroads into a single national company, which was controlled by the government and which enjoyed the support of Americans' competitors—the British S. Pearson and Co.[2]

It was precisely from this time that English and German entrepreneurs became involved in the struggle for Mexican oil. To create a counterweight to the American party, Mexican authorities headed by P. Diaz decided to grant a concession to the British firm of Pearson, Mexican Eagle. The Diaz government did not nourish any special fondness for the English, but rather acted out a simple calculation: "So long as the oil companies are fighting, there will be plenty of money for our purposes."[3] The French newspaper *Le Figaro* observed that "Pearson had obtained certain concessions from the Mexican government, which led him to believe that one day he would become one of the oil kings of the world."[4] Indeed, in just two to three years the English petroleum firm Mexican Eagle grew into a powerful and dangerous rival of the American monopoly. In 1910 it produced 58 percent of all oil in the country.[5] An American newspaper (inspired by the Waters-Pierce Oil Co.) accused Pearson of taking "more out of Mexico than anyone since Cortez."[6]

The bitter competition between the Americans and the English came to be called "the oil war." In March 1910 the American *Oil Investor's Journal* observed that Rockefeller and Pearson were locked in "a cut-rate war."[7] To be sure, in 1909 the two sides tried to reach an amicable settlement, but the effort came to naught. Three years later representatives of the American and English companies made another attempt to resolve their differences in 1912, but once again failed to reach an agreement.

The *Wall Street Journal* noted that the struggle between Rockefeller and Pearson was continuing with undiminished force.[8] The English used the patronage of Mexican authorities to strengthen their position. Unwilling to reconcile themselves to this, the Americans began to support those who opposed the government of Diaz. By this time the internal contradictions in Mexico had become acutely strained, with the revolutionary movement sweeping across the country and dividing the nation into two sides: on the one hand, the landowners with the great latifundia (who were represented by the ruling party), and on the other hand, the rising Mexican bourgeoisie that stood at the head of the Mexican peasant masses. The United States began to support a prominent figure of the liberal democratic groups, F. Madero, who was given the opportunity to form military units on American soil that subsequently became the core of an army. Madero also obtained weapons and ammunition from the United States.

By May 1911 Madero had won the struggle, forcing President P. Diaz to flee the country. However, because the new government could not cope with the rising revolutionary movement in the country, Madero was forced to undertake more extensive social reforms. So far as the oil industry was concerned, Madero declared that he "will encourage American capital investment, but will block access for all trusts and unfair concessions."[9] That kind of program did not, of course, bode well for Standard Oil. The Mexican leader sought to stabilize the situation in the country but, as the Russian envoy justly noted, the American aim was "above all not to permit a strong government in Mexico."[10]

The United States, disillusioned with the Madero government, terminated its support and began to court the opposition. The latter consisted of extreme rightist elements from the latifundia owners, led by V. Huerta. In February 1913 a counter-revolutionary coup d'état occurred; as a result, Madero was slain, his government overthrown, and power given to Huerta, who established a brutal dictatorship over the country. Hoping to domesticate Huerta, an international consortium (which included the American banks of J. P. Morgan, Kuhn Loeb & Co., the Deutsche Bank, Bleichröder's bank, and the Dresden Bank) concluded an agreement in June 1913 granting Huerta a loan for 16 million £ Sterling. According to the terms of the loan, participants of the consortium were given control over Mexican state finances. In addition, the Deutsche Bank expected to acquire a concession for 200 hectares of land that German geologists had already found to be oil-bearing; to exploit this property, the bank planned to create a banking company called Deutsche Petroleum Allgemeine.[11] New forces were thus entering into the oil war, and that in turn led to further complications in the conflict.

The appetites of those making claim to the Mexican oil fields were growing, and the output of oil quickly increased. If in 1910 some 484,000 tons were produced, a year later output was more than three times as much—1,672,000 tons, and by 1913 that production had more than doubled (3,422,000 tons).[12] As a consequence, Mexico rose from seventh to third among world oil producers. Such an explosive growth inflamed passions and intensified stockjobbing.

Another factor contributing to the fever around Mexican oil was the campaign to convert the fuel of naval forces, which spread from England to other countries. "The Mexican oil deposits," wrote the Russian envoy Stalevskii, "attract the attention of many powers and acquire greater significance since contemporary military ships are converting to oil heat (for economic and practical reasons)."[13] The press wrote about this, indicating that "the situation has become seriously complicated by the introduction of oil-burning engines on the military vessels of the royal fleet." "At stake is the fate of England as mistress of the seas," confirmed one of the American journals. "King oil occupies the same position now as did king cotton during our civil war."[14]

Britain's successes in Mexico aroused unconcealed displeasure in Washington. The United States believed that England had violated its rights as master of the Western hemisphere (citing here the Monroe Doctrine) and asked how would England react if the United States intruded into southern Persia, a sphere of British oil interests. The English, however, rejected such arguments. As the newspaper *The Times* asserted, there was no analogy between Mexico and Persia and no sphere of influence in Mexico.[15]

The Huerta regime allowed English capital to reinforce its position in Mexico. For the period from February 1913 to May 1914 alone, British investments in the Mexican oil industry rose by 750,000 £ Sterling. In 1914 the Pearson company produced 60 percent of the oil produced in Mexico. In level of organization and financial capabilities, the British enterprise could compete with the greatest oil trusts of the world. It built oil distillation plants and a fleet of ocean-going oil tankers and opened a commercial representation outside Mexico and Central America. In July 1913 Pearson signed a contract with the British Admiralty to deliver 200,000 tons of liquid fuel. The company opened branches in many Latin American countries—such as Argentina, Brazil, Chili, Uruguay, and Paraguay.[16]

The headquarters of the Pearson firm in London employed 800 people, who worked in five different sections. Each had a different sphere or function: the European Sales Department was in charge of trade in England and continental Europe; the Foreign Sales Department oversaw trade in Latin America, Canada, Australia, Asia, and Africa; the Bunkering Department directed the construction of fueling stations and their activities for the area between New York and Buenos Aires; the Transportation and Buying Departments coordinated the activities of all the organizations in the Pearson concern. To oversee the activities of the various sections, the firm held daily conferences of directors and weekly meetings of the firm's board.[17]

Pearson's company was subjected to vigorous attacks by its American rivals. In the pithy phrase put in circulation by the Americans: Pearson "has taken more out of Mexico than anyone since Cortés." It was not merely a matter of expert propaganda; the American firms also enjoyed the support of the United States Government and its diplomatic service. In particular, the United States began to press for the removal of Huerta, most overtly in a diplomatic note sent to European countries: "If General Huerta does not resign because of the force of circumstances, it will be the duty of the United States to use less peaceful means in order to remove him." True, the American government simultaneously gave European countries a reassurance that "it will not go" further than the "indicated goal," and that it will not begin "to seek any kind of special or exceptional privileges for its citizens either in Mexico or in any other place, but will strive to prove that here, as everywhere else, it is a consistent proponent of 'open doors'."[18] But that was really just a verbal smokescreen. Stalevskii was correct when he suggested in the summer of 1913 that, because Huerta's actions were so repugnant for the United States, the latter was prepared "to lead the country into complete anarchy, heedless of the large capital that Americans had invested in various enterprises," in order "to have the option of intervening in Mexican affairs as occurs in other states of Central America."[19]

The American administration intensified its attacks on the English after President Woodrow Wilson became president in 1912. Wilson declared that it was Pearson who had persuaded the English government to establish diplomatic relations with Huerta. Anglo-American relations grew increasingly strained after L.

Carden—famous for his anti-American sentiments—was appointed British ambassador to Mexico in July 1913. Indeed, his very appointment was attributed to the intrigues of Pearson. The American government made accusations against England, declaring that the Foreign Office had made a secret deal with Pearson to expel the Americans from Mexico—in violation of the Monroe Doctrine.[20]

Wilson's government depicted the British government as little more than Pearson's puppet, although in fact there was no basis for this assertion. To resolve this Anglo-American conflict, in July 1913 President Wilson dispatched Col. E. House to London, where he met with Foreign Secretary E. Grey. The American emissary came away with the impression that "British support of Huerta was neither definite nor final," although Grey remained evasive, preferring to leave the question open for the time being.[21] But in November 1913 Grey's personal secretary, W. Tirrel, arrived in Washington to continue the negotiations. England was forced to retreat. Having been compensated in the question of the levy for passage through the Panama Canal, and not wishing to strain relations with the United States in the face of acute international tensions, the British declared that they had no intention of resorting to political intervention either in the internal affairs of Mexico or in the affairs of any other state of Central or South America. To be sure, it simultaneously made the reservation that it affirms its duty to protect the life and property of English citizens in the event a threat arises. However, England was forced to renounce its support of Huerta, and in essence that meant a renunciation of intervention in Mexican affairs.[22]

In July 1913, amidst parliamentary debates about the conversion of British naval forces to oil fuel, Churchill declared that "the Mexican supplies of oil are abundant and cannot be neglected by the Admiralty."[23] Subsequently, in December 1913, Pearson proposed to the English government to sign a contract, whereby the company Mexican Eagle would be transferred for seven years to the control of the Admiralty, with the obligation to deliver fuel for the navy in exchange for a subsidy of 5 million £ Sterling.

In contrast to the Anglo-Persian Oil Company that was involved in intensive negotiations at this time, the Pearson company was operating at full-steam and could guarantee large-scale deliveries. However, the inevitable complications certain to ensue in Anglo-

American relations made such an agreement impossible. Therefore, amidst the parliamentary debates in June 1914, the government rejected the Mexican option and insisted that the main (though not the sole) source of oil supplies become Persia. The opposition—inspired by vested interests in the oil industry—decried the lack of necessary guarantees for the security of the British position in Persia and urged a repudiation of the agreement with the Anglo-Persian company as an encroachment on the interests of other English oil companies, including those in Mexico. The Foreign Secretary, Grey, declared: "Take the difficulty of protecting these oil wells, and, even at the worst, 150 miles of pipeline from the coast of Persia. Would you rather have oil wells in Mexico?" This was an unequivocal retreat from the earlier British position, especially since Grey embellished this declaration with a respectful and conciliatory comment: "Would you rather have the oils in a country with whom we are on the best relations and with whom we have I would say almost no possible cause of friction, such as the United States?"[25]

Meanwhile, in Mexico, the mass movement against the reactionary landowner regime of Huerta had intensified, making the domestic situation in the country extremely complicated. In July 1913 a Paris conference of French, American, and English bankers appealed to the U.S. Government to begin intervention and to restore order. This proposal was favorably received in Washington and supported enthusiastically by the oil corporations. President Wilson demanded the resignation of Huerta and gave the order to support the followers of V. Carranza, who proclaimed himself a follower of Madero. American petroleum entrepreneurs gave him financial support. Doheny offered Carranza credits for almost 700,000 dollars, promised to provide fuel, and in the event Carranza came to power, vowed to pay the taxes that he had declined to give Huerta. Doheny himself described all this in his testimony to the Senate committee on foreign affairs.[26]

The Mexican dictator Huerta was forced to surrender power to Carranza, whose government was soon granted de facto recognition by the United States. But the revolutionary movement continued to grow in the country, forcing the Carranza government to take measures affecting the interests of foreign (including American) capital. The tax on oil was increased, while the granting of concessions to foreigners became the exclusive prerogative of the central government. Even the question of nationalizing the oil industry was raised. All this, naturally, did not fail to raise alarm in Washington.

As early as the beginning of 1912, when Madero held power, the Russian envoy to Mexico noted that the American government vigilantly follows events in that country, but "cannot make up its mind to intervene, realizing that such a step would arm Mexicans of all parties against it." However, in Stalevskii's opinion, the United States was prepared, if need be, to commence intervention "in view of the enormous interests and capital that the Americans have placed in Mexico."[27]

In the American press and in Congress voices indeed began to clamor for intervention. That demand, furthermore, was supported by representatives of the army and navy. President Wilson noted that influential circles were endeavoring to provoke intervention in Mexico. In April 1914, when Huerta was in power, American troops exploited an insignificant incident to land in Veracruz.[28]

In October 1912 a representative of Standard Oil sent a lengthy note to W. Crozier (representative of the United Committee of the Chiefs of Staff), urging that the Mexican port of Tampico be seized because of its vital importance for American oil fields.[29] After an agreement between the State Department and Navy, the Mexican ports of Veracruz, Tampico, and Puerto Mexico became objects not only of constant visits, but also of the regular moorage of American ships. Although these visits were accompanied by "friendly" reassurances, in fact they were evoked by a desire to show "the strength and power of our navy" in order to produce "a desirable moral effect upon the minds of the local population" and to cause "greater respect to be shown in the future both to this Government and to the citizens of the United States resident in Mexico."[30]

Standard Oil's note to General Crozier stated that the American petroleum holdings have such great significance that the U.S. Army and Navy should seize Tampico and thereby guarantee the security of American oil properties in Mexico. In the summer of 1913, after Woodrow Wilson had become president, Secretary of State W. Bryan expressed doubts about the wisdom of military intervention, which would be tantamount to placing "property rights ahead of human rights," or putting "the dollar above the man."[31]

Representatives of Standard Oil continued to apply pressure, and military circles believed that it is necessary "to do something," and their role in making foreign policy showed an extraordinary growth in this period.[32] It was precisely in the early twentieth century that the diplomatic and military domains began to hold regular consultations, thereby laying the basis for the so-called "military-diplomatic complex." In Mexico the diplomats and military officers worked in close contact, and when, in the spring of 1914, an American naval vessel provoked an incident in Tampico, the American government hastened to exploit it as a ground for military intervention. "Thus the military services," as the American researcher, R. D. Challener, has observed, "spurred by the Standard Oil Company, were drawn to Tampico, the site of the famous naval incident in the spring of 1914, which provoked Wilson's direct military intervention in Mexico."[33] As a result of the American actions, Mexico was faced with a real threat of military occupation. But, fearing that the revolutionary outbursts would intensify, in November 1914 the United States was forced to evacuate Veracruz. However, in March 1916 the American government dispatched a twelve-thousand man contingent of troops to fight units under the peasant leader Pancho Villa, who was active on the American border. Only in February 1917, two months prior to America's entry to World War I, was the U.S. Government forced to recall its troops from Mexico. Their commander, General John Pershing, was appointed as head of the American forces in Western Europe.

Nevertheless, the threat of American military intervention in Mexico remained. The situation became especially tense when the Mexican constitution of 1917 led to decrees imposing limits on the activity of oil companies. The latter vehemently protested the measures taken by Carranza's government and formed the "National Association for the Defense of the Rights of Americans in Mexico," where the leading role belonged to Waters-Pierce Oil Company, the Mexican Petroleum Co., and others. Another byproduct was the formation of the Association of Oil Producers of Mexico. Both these organizations appropriated significant sums for subversive activities against the Mexican Revolution.

Later, a special Senate commission, created to investigate Mexican affairs, sought to free the oil magnates from accusations of preparing intervention against Mexico. However, the evidence it obtained from the interrogation of participants in these events served only to confirm that the petroleum industry were defenders of intervention. The fact that the plans for intervention against Mexico ultimately failed is explained, above all, by the growth of the revolutionary movement and by the international position of the United States—that is, on the eve of its entry into World War I, the country needed to concentrate all its efforts on the European theater of actions.

In the face of the approaching World War and the rising German threat, England virtually capitulated before the United States, and recognized the priority of American claims to Mexico. Some authors have even asserted that the American government made England's renunciation of claims to Mexican oil a condition for its entry into the World War.[34] Although it is doubtful if the question was really posed in those terms, it is quite clear that the Mexican oil problem remained one of the most important issues in Anglo-American relations. It cannot be doubted that for both America and England the German threat held top priority. This was also the case in Mexico, where German banks had been so actively operating. No later than 1910 had a representative of the Deutsche Bank (F. E. Lehner) visited Mexico and examined the possibility of organizing an oil enterprise there. According to the testimony of the Russian envoy in Mexico, this endeavor was inspired by the German navy— it had exerted pressure on a group of German capitalists, who now seriously turned their attention to this enterprise and strove to acquire tracts of oil-bearing land.[35] In 1913 the German company, Deutsche Petroleum, dispatched a geologist (W. Wunstorff), who recommended acquisition of Mexican oil reserves. That recommendation, however, could not be implemented. Finally, not long before the fall of Huerta, the German envoy Paul von Hintze met the dictator and discussed the possibility of Germany acquiring Mexican oil fields. In a report about this meeting (held on May 29, 1914), the envoy wrote:

> Germany is confined within too small a territory.... Germany's natural enemies are England and Russia. Germany wants to colonize and needs oil; he [Huerta—A.F.] is offering Germany 150,000 square kilometers of

land and the oil fields around Tampico, which would be legally taken away from the Americans.[36]

This "generous" offer, however, was made by Huerta in the face of an impending American intervention that left no chance either for Huerta or for the German oil concession. On July 15th the dictator was forced to flee to Europe on board the German cruiser Dresden.

To explain Germany's refusal to establish a base in the Mexican oil industry, the Austrian envoy blamed the solidarity of German banks with Rockefeller's Standard Oil. But such an explanation is not persuasive. The main barrier to German claims to Mexican oil was the determined opposition of the United States and England, which united their forces in the face of the German threat.

NOTES AND REFERENCES

1. *Investigation of Mexican Affairs. Preliminary Report and Hearings of the Committee on the Foreign Relations of the United States Senate* (Washington, 1920), 2, p. 2568.
2. F. Katz, *Deutschland, Diaz und die Mexikanische Revolution* (Berlin, 1964), pp. 61-62.
3. *Investigation of Mexican Affairs,* 2, p. 2636.
4. *Oil War in Mexico. History of a Fortune-Wrecking Fight between Colossal Interests Headed by Mr. H. C. Pierce and Sir W. Pearson,* ed. Petroleum World (New York, 1910), p. 63.
5. Katz, p. 69.
6. A. Jones, p. 71.
7. *Oil War in Mexico,* p. 63.
8. Ibid., p. 65.
9. A. S. Stalevskii to A. A. Neratov (September 29/October 12, 1913). In *AVPR,* f. Kantseliariia, 1913, d. 90, l. 132.
10. Stalevskii to Sazonov (July 29/August 11, 1913). In ibid., l. 53.
11. Katz, p. 69.
12. See Table 5 in Appendix.
13. Stalevskii to Neratov (September 29/October 12, 1913). In *AVPR,* f. Kantseliariia, 1913, d. 90, *1,* p. 132.
14. A. Vagts, *Europa and Amerika unter besonderer Berüksichtigung der Petroleumpolitik* (Berlin, 1928), p. 195.
15. Ibid., p. 134.
16. Ibid., p. 196; report from the U.S. intelligence agent in Tampico (February 24, 1913). In United States National Archives, U.S. Office of Naval Intelligence, RG 38, Box 38, File 2647.

17. Jones, pp. 69-70.
18. Note from the U.S. Government to the Russian Government (November 13/26, 1913). In *AVPR,* f. Kantseliariia, 1913 g. d. 90, ll. 157-158.
19. Stalevskii to S. D. Sazonov (July 28/August 11, 1913). In *AVPR, 1,* p. 54.
20. Jones, p. 71-72.
21. *The Intimate Papers of Colonel House,* arranged as narrative by Charles Seymour (Boston, New York, 1926), *1,* p. 195-197.
22. Ibid., *1,* pp. 198-201; *Blue Book. United States No. 1: Oil Properties and Mining Rights in Mexico* (London, 1914), pp. 1-3.
23. *Parliamentary Debates,* July 17, 1913, *55,* p. 1477.
24. Jones, p. 76.
25. *Parliamentary Debates,* June 17, 1914, *63,* p. 1185.
26. *Investigation of Mexican Affairs, 1,* pp. 278-279.
27. Stalevskii to Sazonov (February 25/March 9, 1912). In *AVPR,* f. Kantseliariia, 1912 g., d. 85, ll. 10-11.
28. L. C. Gardner, *Wilson and Revolutions, 1913-1921* (Washington, 1976), pp. 17-18.
29. R. D. Challener, *Admirals, Generals and American Foreign Policy, 1898-1914* (Princeton, 1973), p. 357.
30. Ibid., p. 353.
31. Ibid., pp. 356-357, 382-383.
32. Ibid., ch.1.
33. Ibid., p. 357.
34. K. Gofman, *Neftianania politika; anglo-saksonskii imperializm* (Leningrad, 1930), p. 63.
35. Stalevskii to Neratov (September 28/October 12, 1913). In *AVPR,* f. Kantseliariia, 1913 g., d. 90, l. 133.
36. Katz, *The Secret War in Mexico, Europe, the United States and the Mexican Revolution* (Chicago, 1981), p. 240.

Chapter VII

The Vicissitudes of the Russian Oil Business

After the turn of the century and especially in the immediate prewar years, the development of the petroleum business in Russia took on a highly contradictory character. On the one hand, it experienced the growth and consolidation of huge modern corporative conglomerates and also a stronger influx of foreign capital after the discovery of new oil reserves. On the other hand, as a consequence of the economic crisis in 1900-1903, the strike movement in 1903-1904, and the revolutionary upheaval of 1905-1907, there was a decline in both production and trade. Thereafter, to be sure, the industry began a recovery, but it did so at a relatively slow tempo. Russia's relative share of world output fell from 53 percent (1901) to 16 percent (1914). The production and trade of petroleum products also fell in absolute terms;[1] the country experienced a shortage of fuel, leading to "an oil famine."

That whole period of crisis—from industrial depression to revolution—profoundly shattered hopes for a miracle of economic reform. The debris of these eventful years entombed Witte's program once and for all. In Nobel's words, "two special circumstances"—military defeat in the Far East and revolution at home—had a catastrophic impact on the Russian oil business. Both in various meetings and also in memoranda to the tsarist government, Nobel complained of the calamitous condition of oil production and export, urged the government "to end the war at any cost," and demanded vigorous measures to quell the strikes and revolution.[2] The Baku plants, the railways, and Batumi port were all idle; oil fields were engulfed in flame. "The strained relations" between entrepreneurs and workers, complained the British consul in Batumi, often played havoc with the flow of plant operation, "upon which the regular rotation of trade so much depends."[3]

The oil business brought in enormous profits to the entrepreneurs, but the condition of the workers was onerous. The writer Maksim Gor'kii had visited Baku as far back as the 1890s and observed that the oil fields remain in his memory as "an ingenious representation of pure hell."[4] It is therefore not surprising that the oil capital of Baku and the petroleum gateways to Russia of Batumi were fated to become important arenas of revolutionary activity. But what transpired there was only an echo of the revolutionary movement sweeping the entire country, with the epicenters located in Moscow and St. Petersburg.

Because so much has been written and published about the Revolution of 1905-1907, we shall limit our discussion here to quotations from the previously unpublished letters of L. L. Pershke (chief of the Office for Excise Tax Assessments for the Trans-Caucasus Region), one of the most well-informed people in the area of Russian oil business. Pershke's letters to his wife, which were intercepted and sent to the Department of Police (in whose archive they are now preserved), describe the situation on the eve of the revolution. At the end of 1904 Pershke was in the capital and wrote his wife in Tiflis:

> There is general ferment in St. Petersburg; the revolution is constantly being prepared; in any event, an open struggle with the old regime is be waged in all strata of society. They demand reforms, which the government will not be able to ignore. The situation is extremely tense and the authorities have lost their heads. Each day meetings are being held everywhere, not only gatherings of youth, but also of various professions, and demands are being formulated in written documents.[5]

Similar events were unfolding in Baku and Batumi, and were intimately linked to what was happening in St. Petersburg.

The magnates of Russian oil—Nobel, the Rothschilds, Mantashoff, and others—demanded that the government take urgent measures to restore law and order. The development of the revolutionary movement provoked a mood of panic among British entrepreneurs, who inundated Petersburg with letters and telegrams demanding "proper and adequate protection ... [of] the lives and property of the foreigners in Baku and other parts of the Caucasus." The English ambassador in St. Petersburg, C. Hardinge, sent notes to the Ministry of Foreign Affairs on two separate occasions—in August and September 1905.[6] Calls to reestablish "law and order" also resounded at the meetings of the alliance of entrepreneurs—the Council of the Congresses of Baku Oil Industrialists. These leading oilmen expressed doubts about the capacity of the government to deal with the country's political crisis and made proposals regarding the necessity of reform. "The social development of the Russian state has outgrown its political jacket," wrote the paper *Neftianoe delo* ["Oil Business"], "and the inevitable transition to new forms of political order has arrived."[7] Nobel said approximately the same thing: at the very height of the revolution, when he was asked "What measures, in your opinion, could prevent the strikes?" Nobel replied: "A change in the state system."[8] But of course the revolution did not culminate in that kind of fundamental restructuring of autocracy and its relationship to society, although it did succeed in throwing a fright into the ruling classes.

At the same time, it should be said that, despite the tumult and confusion of the time, it was precisely in these years that decisive steps were taken toward the further monopolization of the Russian oil industry. As early as 1903 Nobel and the Rothschilds had signed an agreement to create a cartel called "Nobmazut" for the domestic petroleum trade. In 1905-1907 they added the association of British entrepreneurs Shiboleum (a combination of the two companies, Shibaeff and Oleum) and the Mantashoff group, who renounced domestic petroleum trade in favor of Nobmazut.[9] These agreements, which lasted for several years, were strictly confidential; when Nobel informed his Baku office of their signing, he admonished his subordinates that "these agreements are to be kept secret."[10]

As a result of the agreements, their participants—even amidst conditions of falling production and declining trade—were able to bring in relatively high profits because of the monopolistic prices on the market. Thus, at the very time that smaller firms suffered losses, the large corporations found a way out of their difficult situation. In July 1906 Nobel complained that Russia was forced to export "products from the processing of crude oil—kerosene and lubricants, ... leaving in our country the 'oil chaff'—residual fuel oil, which serves us as fuel."[11] Significantly, however, Nobel and his allies had a vested interest in sustaining the "oil famine." In 1904, from an output of 9.2 million tons of oil, the income from its sale amounted to 90 million rubles; in 1906, the income from a smaller production (6.9 million tons) was 115 million rubles.[12] Judging from the reports of the Nobel Company, in spite of reverses caused by the crisis, it continued to operate at a profit. Other concerns were also making profits, even if the fall-off in business and the general political turmoil could not fail to evoke alarm among them.

THE SALE OF THE ROTHSCHILDS' ENTERPRISES

It was under these circumstances that negotiations were begun to sell the Rothschilds' enterprises, culminating in a sales agreement with Royal Dutch Shell. Deterding's motive, as already indicated, was a desire to expand, at any cost, the production base of Royal Dutch Shell. The attention of the powerful Anglo-Dutch trust was attracted to Russia by the growing interest in Russian oil securities on the London Stock Exchange. The cause of the new oil fever was the discovery of new reserves in the northern Caucasus (in the area around Maikop), where several British petroleum companies were formed immediately. Concurrently, English capital expanded its investments in Baku, Groznyi, and the Ural-Caspian area. The total sum of British investments of the ensuing years—amounting to 180 million rubles—would double that of the 1890s. A new wave of British investments, accompanied by a speculative fever, was actively supported by the Foreign Office.[13]

As it began negotiations to enter the oil business in Russia, Royal Dutch Shell intended to exploit the favorable tenor in Anglo-Russian political relations after the agreement of 1907, which created excellent prospects for British entry to the petroleum industry in Russia. Speaking of Royal Dutch Shell's efforts to become established in Russia, one of Deterding's closest associates, A. J. Cohen, emphasized that "it is very desirable to set up a British company, since we should then enjoy the support of British diplomacy in Russia."[14]

It was also quite clear that the tsarist government would have to pay heed to English claims. At the height of the Maikop fever, an agent for the Ministry of Finance in London warned his superiors in St. Petersburg that even a delay in issuing permits to British companies could give ground "to diplomatic representations from the English government."[15] This alarm was entirely groundless, however: the Russian government not only raised no objections to the English claims, but even encouraged them. The minister of finance, V. I. Kokovtsov, declared that the majority of members of the Council of Ministers (including himself) "in no way will adopt a viewpoint that violates the interests of existing foreign companies or in general that is hostile to English capital."[16]

Although Witte, whose name was intimately associated with a favorable view of admitting foreign capital in the 1890s, had fallen into disgrace and was currently in retirement, his policy toward foreign capital continued. The Council of Ministers adopted a special decree reaffirming the decision of a Special Conference in 1898 on the rights of foreigners to participate in the Russian oil business.[17]

In May 1909 Lane informed Nobel that Deterding "is planning, it seems, to become involved in the Russian oil industry." True, he declined an offer to purchase Mantashoff's stock, preferring instead to do business, as before, with his "Paris friends"—Lane and the Rothschilds.[18] In the words of F. Gerretson, the historian of the Anglo-Dutch company, the question of an agreement between Deterding and the Rothschilds "was in the offing since amalgamation"—that is, ever since Royal Dutch and Shell merged in 1907.[19] Moreover, it can be said that this question was already on the agenda at the turn of the century, when, during the negotiations to create the concern "Asiatic," Deterding had emphasized that one must act "together with the Rothschilds."

Thus the idea of an agreement did not suddenly appear, but developed as a result of prolonged joint activities. It was undoubtedly reinforced by the political compatibility of the Anglo-Dutch oil trust and the Parisian bankers. Nevertheless, when, in the spring of 1909, Deterding initiated discussions with Lane on this subject, the latter deemed such a deal highly improbable. Doubting that the Rothschilds would want "to have a connection to this," he "refused to make them a proposal."[20]

The coming months, however, showed that Lane's doubts were unfounded. The Rothschilds expressed a willingness to begin negotiations and began the discussion with the possible sale of the firm "Russian Standard," which was the largest enterprise in the Groznyi oil district and the second largest in the whole country. The plants of Russian Standard were experiencing economic difficulties and were in acute need of financial assistance. They obtained support from Royal Dutch Shell, which purchased Russian Standard and also two other firms in Groznyi. After some reorganization, the company "New Russian Standard"—as Deterding renamed the Groznyi firm—raised its level of activity and began to yield a profit. This success was augmented by the formation of the Ural-Caspian Oil Company to exploit the promising oil fields of the Emba district.

It should be emphasized, however, that the main objects of Royal Dutch Shell's interest were the Baku properties of the Rothschilds—the Caspian and Black Sea Co. and Mazut, which (along with the Nobel firm) were the leading firms for oil production and trade in Russia. Deterding approached the question of acquiring these properties as a result of complex and prolonged negotiations under Lane's mediation. "Everyone here was able to value the important role that you played in the negotiations," Aron wrote Lane, after the agreement for Deterding to purchase the Rothschilds' Baku enterprises had been initialed in November 1911.[21] The final terms of the purchase agreement were eventually worked out, making Royal Dutch Shell the sole owner of all the Rothschilds' enterprises in Russia.

News that the Rothschilds' properties had been sold caused a sensation: it was the largest such business deal in the entire history of the Russian oil industry. At the same time, however, it was not all that extraordinary. A study of documents on the negotiations with Deterding in the Rothschilds' archives (by the Soviet historian, V. I. Bovykin) has shown that there was really nothing so unusual here. Although conducted in the strictest secrecy, the negotiations, in themselves, were quite routine.[22]

What impelled the Rothschilds to sell their enterprises to Deterding? Bovykin has observed that the archival documents "do not directly answer" this question. At the same time, he came to the conclusion that "the Rothschilds did not plan a general withdrawal from Russia," that their decision to abandon the oil business "was not the result of their own initiative," but rather was "foisted on them in the course of the negotiations with Deterding."[23]

Insofar as Bovykin's interpretation lacks documentary support, it inevitably raises some doubts, all the more since materials from another source—the Nobel archives—support precisely the contrary interpretation.

The crises besetting the Baku oil industry, as a result of the Revolution of 1905-1907, had a major impact on the state of the Rothschild enterprises. Their chief rival, Nobel, was also affected. But the Nobel firm's condition was more secure both in economic terms and in other respects. Occupying the commanding position in the Russian oil industry, Nobel strove to subordinate the Rothschilds and to force them to agree to coordinate their operations. In December 1908 he posed the following question to the Parisian bankers: do they intend "to act together with him" or do they plan "to assume another position?"[24] On another occasion, a member of the Nobel board, E. K. Grube, sent Aron a letter with the conditions, under which the Rothschilds should "constantly follow us, even when they should not agree with what we have in mind."[25] At issue was a particular case (joint trade of lubricants in Europe), but here too Nobel was adamant in insisting that he hold the dominant position.

The relations between Nobel and the Rothschilds, in general, did not follow the optimal path of development. In a letter to I. O. Olsen, the manager of the Nobel firm, E. K. Grube remarked that "the Rothschilds are becoming weary" and that therefore their enterprises—the Caspian and Black-Sea Co. and Mazut—could be sold in part or in their entirety to Deterding, who was intending to purchase "these two companies or one of them in order to be able to negotiate with Standard Oil regarding an association encompassing the whole globe." Grube also insisted that "all this would perhaps be pure fantasy" had not Deterding already expressed a desire to buy other Baku firms and proposed to Nobel "to acquire together the Mantashoff company."[26] Hence it was as though Deterding's deal with the Rothschilds had been predetermined in advance.

As already noted, the two parties did reach an agreement. The Rothschilds sold their properties to Deterding for a huge sum—approximately 35 million rubles, which, according to Gerretson's assessment, represented 50 percent over their real value.[27] After the agreement had been initialed, the head of Royal Dutch Shell began to delay signing the agreement. To obtain prompt implementation of the agreement, the Rothschilds decided to give up 1 million rubles in the final assessment of the property of the Caspian and Black Sea Company. They also agreed that the allotted sums be paid not in cash, but in the stocks of Royal Dutch Shell.

The terms of the agreement eloquently testify to the mutual interest of both parties. Deterding strove to gain control of Russia's oil riches and obtained them. The Rothschilds agreed to give up their holdings in Russia, but retained their investments in the oil business by acquiring 20 percent of the stock in the Royal Dutch Shell trust (with the stipulation that it not be posted on the stock market before January 1914). At the same time, the Rothschilds remained stockholders of Russian enterprises, in particular, of Mazut, right until 1916, as the notations in company records indicate.[28]

The purchase of the Rothschilds' enterprises by the Anglo-Dutch trust in 1912 marked an important stage in the international struggle over oil. Bovykin argues that this event "reflects the key objective processes at the beginning of the twentieth century in the formation of mature forms of finance capital and, on this basis, a change in the distribution of forces in the struggle for the world market."[29] But it must be said that Deterding's deal with the Rothschilds could have only occurred under the condition of a consolidation of Anglo-French economic and political relations within the context of the growing role of the Entente.

However, one might well ask why the French, who had previously had closer ties with Russia than the English, should now suddenly renounce their share of the Russian petroleum business. After all, they had attained their position only at considerable difficulty. The suggestion that they withdrew simply because of Deterding's pressure does not appear very cogent. It is far more likely that the Rothschilds had lost their earlier passion for the oil business in Russia. This was not only because of economic losses and difficulties from a competitive struggle, but also because Russia's political development—accompanied by all the strikes and social turmoil—did not inspire much optimism. The military defeat against Japan and especially the Revolution of 1905-1907 shattered their confidence in the tsarist regime. As the minister of finance, V. N. Kokovtsov, observed, "the tragic events in Baku (the focal point of our oil industry) made the strongest impression abroad, which was extremely unfavorable for our credit."[30] Business circles began to sense the instability and fragility of tsarist authority—phenomena that could hardly have gone unnoticed on Rue Laffitte. One more reason could have provoked the Rothschilds' decision to get rid of their holdings in Russia—the growing importance of the Jewish question, which formed a stumbling block in relations between the Tsarist government and Jewish financiers in the West.

But all these considerations must raise a further question: why did the clever, perspicacious Deterding throw caution to the winds and purchase the Russian enterprises? The explanation apparently must be sought in the fact that Royal Dutch Shell, as a petroleum firm of incomparably larger scale, aspired to world hegemony and countenanced a higher degree of risk than even such major entrepreneurs like the Rothschilds could allow themselves.

In any event, with its acquisition of the Rothschild plants, Royal Dutch Shell joined the ranks of the most powerful oil producers in Russia, their capital comprising more than 60 million rubles by 1914. Her possessions included the Caspian and Black Sea Co. (10 million rubles), Mazut (12 million rubles), three firms in Groznyi (19 million rubles), the Ural-Caspian Co. (7 million rubles), the Schibaeff company and several other smaller firms.[31] After purchasing the Rothschilds' enterprises, Deterding carried out a thorough-going reorganization, including changes in managers. He lured over two Nobel directors, E. K. Grube and G. P. Eklund, and put them on his company's board of directors. Royal Dutch Shell also sought to reach an agreement with the Nobel company, but in vain. Nobel did not wish to share his influence in the governmental circles of Russia; as earlier, he continued to rely upon their support and his close ties with the banks.

Despite the serious deterioration in the condition of the oil business in Russia, the share of Nobel's company in production and trade remained virtually unchanged, declining by just 1 percent for the decade of 1901-1910 (from 11.9 to 10.9 percent). To compensate for the shortage of crude oil, Nobel bought up a significant volume of oil from his firm's allies, which he then processed in his refineries and sold on the domestic and international markets as his own. Nobel's plans for 1911 foresaw that output in his own oil fields would constitute approximately 1 million tons, and that approximately an equal volume would be purchased from other firms. And in 1914, compared to the 1.1 million tons of its own production, the firm proposed to buy another 800,000 tons.[32]

By purchasing large quantities of oil from allied firms, between 1901 and 1910 Nobel succeeded in significantly expanding its production of kerosene and its domestic sales of kerosene and fuel oil. Measured in terms of Nobel's share of the Russian oil business, its production of kerosene rose from 15.2 to 27.1 percent, while its kerosene sales increased from 40 to 51.4 percent and fuel oil sales edged up from 27.8 to 32.7 percent. At the same time, Nobel's share of kerosene exports nearly doubled—from 16.4 to 31.4 percent. It is therefore not surprising that, despite the substantial reduction in profits and dividends in certain years, on the whole its income steadily increased.[33]

The key to Nobel's financial successes was a low per unit cost of oil production at its own enterprises and the relatively moderate purchase price of additional oil from clients—the petty and middling oil producers. A graphic picture of how this worked in practice is to be found in the dynamics of prices on kerosene.

In the prewar years, as at the turn of the century, Nobel was frequently accused of sacrificing the interests of the Russian petroleum business by maintaining relations with the American trust, Standard Oil. This notion was so deeply implanted that a 1911 report to U.S. President W. H. Taft on American investments in Russia (compiled, at his request, by experts in the Commerce Department) consigned the Nobel company to the category of American enterprises.[35] That of course was a gross distortion: although Nobel continued to support contractual relations with Rockefeller's company for the sale of oil products on the world market, neither Standard Oil nor its partners owned stock in the company.

Indeed, the Nobel corporation remained an independent firm, and although it did maintain close contacts with a number of Russian and foreign banks, it was self-financing. As the chairman of the board of the Volga-Kama Bank and board member of the Azov-Don Bank (the largest financial institutions operating inside Russia), Nobel attained influence over a whole series of oil companies, which, in some degree, proved dependent upon him. Nobel's firm continued to receive money from the German bank "Disconto Gesellschaft," but maintained its independence by using a variety of devices. For example, to sustain his firm's autonomy, in 1910 Nobel took a loan from the Swedish "Enschilde" Bank so that he could pay off a debt of 4 million marks to the Disconto Gesellschaft.[36]

During the prewar industrial boom in Russia (1909-1913), Nobel increased its ties to Russian banks, thereby making use of their heightened interest in Russian enterprises. In 1912 the company's capital was increased from 15 to 30 million rubles. This operation was conducted by the Berlin Disconto Gesellschaft through an agreement with Nobel.[37] The latter began to place all the new stocks (set at 15 million rubles) on the stock exchange, but surrendered a significant part of the stock to Russian banks. The syndicate established by the Disconto Gesellschaft included, as subparticipants, the leading banks of St. Petersburg and Moscow—the Russian-Asiatic Bank, the Volga-Kama Bank, the Azov-Don Bank, the St. Petersburg International Commercial Bank, the Discount and Loan Bank, the Russian Bank for Foreign Trade, the Commercial-Industrial Bank, the Siberian Bank, Junker and Co., and others.[38] A syndicate, headed by the Russian-Asiatic Bank, was formed to place the Nobel securities on the Russian Stock Exchange and assumed responsibility for stocks worth 1,250,000 rubles.[39]

During a general meeting of stockholders of the Nobel company in May, 1914, representatives of Russian banks held 18 million rubles out of the company's total 30 million rubles in stocks.[40] Quotations on the Nobel securities ran high, and banks eagerly acquired them. The general sum of the company's stock reached the colossal amount of 78 million rubles (30 million rubles in stocks; 16.3 million rubles in bonds; and 32 million rubles in a reserve fund).[41] The total capital of firms under Nobel's sphere of influence comprised 10 million rubles in 1914.[42] In addition, from the end of the nineteenth century Nobel owned a machine-building plant in St. Petersburg, which was engaged in the production of diesel motors, steam boilers, fuel-oil burners, and other equipment. In 1913 Nobel invested 1 million rubles to form a shipbuilding firm called Noblessner, which monopolized the construction of submarines in Russia. This was a whole empire.[43]

The Nobel concern was stronger than ever. But that does not mean that it was immune to competition or attempts to undermine its dominant position. An especially serious threat in this respect was presented in 1912 with the formation of the Russian General Oil Corporation, a powerful holding company that combined numerous enterprises for the production, refinement and sale of oil. The total amount of its capital amounted to 120 million rubles. Although the company was registered in London as a British firm (under the name Russian General Oil Corporation) in order, as its founders explained, to obtain better access to the European money markets, it was really a Russian enterprise. The company attracted the participation of the greatest Russian banks, including the Russo-Asiatic Bank, St. Petersburg International Commercial Bank, the Discount and Loan Bank, the Commercial Industrial Bank, the Siberian Bank and the Russian Bank for Foreign Trade. It is indicative that neither the Volga-Kama nor the Azov-Don Banks—close allies of Nobel—chose to participate in the Russian General Company, regarding its creation has a hostile act against Nobel.[44]

By the end of the nineteenth century the idea of creating a "third force" as a counterweight to Nobel and the Rothschilds had already appeared. It emerged during the negotiations between the St. Petersburg International Commercial Bank and the German financial group that was seeking to establish a base in the Russian petroleum industry.[45]"There now remain two solutions," Rothstein wrote to Warburg, "either [acquiesce in] the initiative of Rothschild and Nobel (and these gentlemen need neither advice nor assistance), or organize an independent company ... In my opinion, a third opinion does not exist." Realization of this project demanded, at a minimum, fifteen to twenty million Marks (approximately 10 million rubles). Therefore Rothstein sarcastically observed that "everything is ready except the money, but that is the main thing."[46] As already noted, the German project failed for many reasons, not only because of finances. The idea of a "third force" had long stood on the agenda, and for more serious reasons, viz., as a consequence of the existence of the Mantashoff group. The formation of the Russian General Corporation realized this idea, with the direct participation of that group (whose core consisted of the firms of Mantashoff, Gukasov, Lianozoff [Lianozov] and the company Neft', which combined small firms around it).[47]

In the words of the well-known banker E. Agahd, the Russian General Oil Corporation was a "banking combination."[48] Its main goal consisted in mobilizing the financial resources of the European stock exchange. In Paris and London, branches of the Banking House of O. A. Rosenburg were opened; along with the French bank of L. Dreyfus and the Belgian baron Margulis, it became the company's main agent in London, Paris, Brussels, and Amsterdam. Their foreign offices represented the interests of Russian banks, which participated in the organization of the company and the subsequent formation of various syndicates to increase the capital of the constituent companies. In the course of 1912-1913 the three main participants in the Russian General Oil Corporation—the firms of Mantashoff, Lianozoff, and Neft'—increased their capital, altogether, from 24.5 million rubles to 76.8 million rubles.[49]

At the end of 1909, the board of Mantashoff's company complained of business difficulties, and petitioned the government for permission to reduce its fixed capital by 50 percent, from 22 to 11 million rubles.[50] In early 1913, it declared that the "enterprise operates almost entirely on borrowed means," and requested permission to increase its capital by another 9 million rubles (i.e., to 20 million rub.)[51] Still more impressive was the issue of new stock by the Lianozoff Co. (increasing its capital from 8 to 30 million rub.) and Neft' (an increase from 5.5 million rub. to 27.5 million).[52]

In the course of the first year and half that the Russian General Oil Corporation existed, the collective efforts of the Banking Houses of Rosenburg, Dreyfus, and Margulis succeeded in realizing the stock of the London-registered "Russian General Oil Corporation" for a sum of 2.5 million £ Sterling (approximately 23.7 million rubles). This generated confidence in the company and enabled it to undertake further operations to expand its capital. Large sums of money were received, and these were in turn multiplied through supplementary loans (with the stock as security). In addition, the pure profit that the banks obtained from the operations with the company's securities amounted to 3.6 million rubles.[53]

In addition to the oil fever, generated by the discovery of new oil reserves, the company's success owed much to the favorable economic situation of the prewar industrial boom and the expanded authority of Russian banks. After uniting a large number of firms around itself, the Russian General Oil Corporation—in the scale of operations—surpassed the leading oil industrial groups of Nobel and Royal Dutch Shell:

Oil Production in Russia[54]
(in millions of tons)

Company	*1913*	*1914*
Russian General Oil Corporation	2.1	1.9
Nobel Brothers	1.3	1.2
Royal Dutch Shell	1.3	1.5
Other firms	4.4	4.3
Total	9.1	8.9

Thus the three leading oil producers accounted for 52 percent of all output. With respect to the market, the above groups controlled approximately 75 percent of all commercial operations, which, as seen in the Nobel case, was achieved by purchasing substantial quantities of oil from middling and smaller firms. Although the latter continued to exist on a more or less independent basis, the task of obtaining credit from Russian and foreign industrial and financial groups remained a prerequisite for their existence. From the late nineteenth century, this practice was conducted in a more demonstrative form by the Rothschilds' company, whose methods were successfully adopted in the future by Nobel, Mantashoff and other magnates of Russian oil.

Swept away by its own success, the Russian General Oil Corporation decided to throw down the gauntlet to the powerful conglomerate of Nobel: it formed a syndicate to buy up his stock. At the head of the syndicate was the Russo-Asiatic Bank, which by then had become the largest Russian bank in the volume of its capital as well as the scale of operations on the Russian and foreign money markets. In the period from April to December 1914 the syndicate acquired a large bloc of Nobel stock on the Berlin and St. Petersburg stock exchanges. Such an attempt was undertaken later by one other syndicate, again headed by the Russo-Asiatic Bank. Both efforts, however, miscarried.[55]

From the very beginning, the activity of the Russian General Oil Company was dominated by a speculative tendency that undermined its position. The enormous returns to the company partners were used not to expand oil production, but to purchase new firms and means of transportation. Such operations became an end in themselves and did not provide sufficient support for the firm's production base. The inevitable result was a decline in the firm's stock. Although Russian banks provided powerful financial support, the attempt to guarantee a constant influx of foreign capital proved futile. The business press sounded the alarm that the company's stock was being sold at inflated prices.[56] Notwithstanding the fact that, at the beginning of 1914, the company's board of directors issued a public declaration proclaiming the "splendid condition of its affairs," in a transparent effort to persuade investors that they "can be confident that their stocks are secure and highly profitable," a powerful blow had been dealt to the company's reputation.[57] In Paris and London, it became increasingly difficult to sell the company's securities. The board of the Russo-Asiatic Bank learned that even Crèdit Lyonnais—which traditionally supported Russian enterprises on the French stock exchange—was "advising its clients to get rid of the stock of the Russian General Corporation."[58] To make things worse, the Nobel company announced in 1914 that it would pay its stockholders dividends of 26 percent, while the Russian General Oil Corporation paid an average of 6 percent and some of its participants (including Mantashoff and Lianozoff) could pay nothing at all.

The boom turned into bust. The epic came to an end in 1916 and early 1917, when the Russian General Oil Corporation was forced to make an agreement with Nobel for an exchange of stock and an adjustment of their interests in a fashion that favored Nobel.[59] Despite the powerful support of the leading Russian banks, which participated in the creation of the firm and actively supported its efforts to establish hegemony over the petroleum business in Russia, this adventure collapsed with crash. As before, Nobel preserved its position as the leading oil corporation of Russia, which it had substantially reinforced by using the favorable market conditions of World War I.

NOTES AND REFERENCES

1. See Table 5 in the Appendix.
2. Memorandum from Nobel (July 12/25, 1906), "K voprosu ob eksporte nefteproduktov." In *TsGIA SSSR,* f. 23, op. 17, d. 532, ll. 351-352; B. A. Romanov, *Ocherki diplomaticheskoi istorii russko-iaponskoi voiny. 1895-1907* (Moscow-Leningrad, 1956), p. 339; A. A. Fursenko, *Neftianye tresty i mirovaia politika (1884 gg. - 1918 g.)* (Moscow-Leningrad, 1965), p. 255.
3. Stevens' report for 1904 (attached to the communication of April 13, 1905). In *British Diplomatic and Consular Reports,* A.S., N 3366, p. 3.
4. M. Gor'kii, *Sochineniia, 17,* p. 113.
5. Excerpt from a letter ("obtained by means of an agent") from L. L. Pershke to O. A. Pershke (November 24/December 7, 1904). In *TsGAOR,* f. Department politsii, Osobyi otdel, 1904, g., d. 2200, l. 3.
6. Parker to Piz (May 9/22, 1905), in the Public Record Office, Foreign Office, 65/1720; Fursenko, *Neftianye tresty,* pp. 241-242.
7. *Neftianoe delo,* November 15, 1905.
8. *Revoliutsiia 1905-1907 gg. v Rossii. Dokumenty i Materialy,* ch. 2 (Moscow 1955), p. 286.
9. P. V. Volobuev, "Iz istorii monopolizatsii neftianoi promyshlennosti dorevoliutsionnoi Rossii (1903-1914 gg.)," *Istoricheskie zapiski, 52,* (1955) pp. 89-94; *MKNPR,* pp. 323-331, 426-428, 404-405.
10. Pravlenie "Br. Nobel'" to Baku Office (April 9/22, 1905). In *TsGIA AzSSR,* f. 798, op. 1, d. 341, l. 67.
11. Memorandum of Nobel (July 12/25, 1906). In *TsGIA SSSR,* f. 23, op. 17, d. 532, l. 326.
12. Gerretson, *3,* p. 152.
13. Smith (consul general in Odessa) to the Foriegn Office (June 27/July 9, 1910); Memorandum "Maikop Association LTD" (September 2/15, 1910). In Public Records Office, Foreign Office, 368/453; Iu. I. Vintser, *Angliiskie kapitalovlozheniia za granitsei* (Moscow, 1960), p. 34.
14. Gerretson, 4, p. 133.
15. M. V. Rutkovskii to Kokovtsov (December 23, 1920/January 5, 1911). In *MKNPR,* p. 498.
16. Kokovtsov to Sazonov (August 11/23, 1910). In *AVPR,* f. Persidskii stol, d. 4452, l. 10.
17. *MKNPR,* p. 724.
18. Lane to E. Grube (May 6/19, 1909). In *MKNPR,* p. 447.
19. Gerretson, *3,* 270.
20. Lane to Grube (May 19, 1909). In *MKNPR,* p. 447.
21. V. I. Bovykin, "Rossiiskaia neft' i Rotschil'dy," *Voprosy istorii* (1978), *4,* p. 38.
22. Ibid., pp. 32-40.
23. Ibid., p. 40.
24. Grube to Aron (December 10/23, 1908). In *TsGIA SSSR,* f. 1458, op. 1, d. 1747, l. 133.

25. Grube to H.A. Olsen (April 30/March 12, 1909). In *TsGIA SSSR,* f. 1458, op. 1, d. 1747, l. 133.

26. Grube to Olsen (October 16/29, 1909). In ibid., l. 276.

27. Gerretson, *4,* p. 136.

28. Board of Mazut to the Ministry of Finance (1916). In *TsGIA SSSR,* f. 1450, op. 1, d. 27, l. 444.

29. Bovykin, "Rossiiskaia neft'," p. 41.

30. Quoted in *Istoriia Azerbaidzhana,* Baku, 1960 *2,* p. 561.

31. L. Eventov, *Inostrannyi kapital v neftianoi promyshlennosti Rossii, 1874-1917* gg. (Moscow-Leningrad, 1925), p. 92.

32. Tovarnaia programma 'Br. Nobel'" na 1911 i 1914 gg. (TsGIA SSSR, f. 1458, op. 1, d. 257, l. 46, 116).

33. See Table 3 in the Appendix; S. and L. Pershke, *Russkaia neftianania promyslennost', ee razvitie i sovremennoe polozhenie v statisticheskikh dannykh* (Tiflis, 1913), pp. 191-193; D'iakonova, pp. 125-126.

34. See Table 4 in the appendix.

35. W. H. Taft Papers, Series 6, f. 205 (Library of Congress, Washington, D.C.)

36. D'iakonova, pp. 114-115.

37. This agreement was reached as a result of a resolution by a general meeting of the stockholders of "Nobel Brothers" on June 18/July 1, 1910, which was then formally approved by the emperor on November 22/December 5, 1911. The text of the agreement between the "Nobel Brothers" and Disconto Gesellschaft, signed the very same day, is in *TsGIA SSSR,* f. 630, op. 2, d. 297, 11. 7-8.

38. Syndicate agreement of December 18/31, 1911. In ibid., ll. 5-6.

39. *MKNPR,* p. 727.

40. List of stockholders of the Nobel company at a general meeting (May 13/26, 1914). In *TsGIA SSSR,* f. 595, op. 2, d. 203, ll. 203, 11, 55-57.

41. Eventov, p. 94.

42. Ibid., p. 96.

43. A list of stockholders of Noblessner at general meetings on January 8/21, 1913 and May 29/June 11, 1915. In *TsGIA SSSR,* f. 23, op. 12, d. 1588, ll. 53, 103-104.

44. *MKNPR,* pp. 7310-7331.

45. D'iakonova, pp. 128-129.

46. Rothstein to M. Warburg (October 31/November 12, 1898). In *TsGIA SSSR,* f. 626, op. 1, d. 1388, *1,* p. 487.

47. *MKNPR,* pp. 730-731.

48. E. Agahd, *Grossbanken und Weltmarkt* (Berlin, 1914), p. 99.

49. Eventov, pp. 80-84.

50. Doklad Ministerstva torgovli i promyslennosti (December 15/28, 1909). In *TsGIA SSSR,* f. 23, op. 24, d. 620, ll. 238-240.

51. Pravlenie "A. I. Mantashoff i Co."—Ministerstvu torgovli i promyslennosti (February 28/March 13, 1913). In ibid., l. 253.

52. Eventov, pp. 80-84.

53. Agahd, p. 186.

54. Tablitsa dobychi nefti po firmam (February 15/28, 1915). In (*TsGIA SSSR,*. f. 1458, op. 1, d. 260, ll. 27-28.)

55. *MKNPR,* pp. 738-739.

56. Letter from one of the holders of securities of Russian oil companies to the editors of the French journal, *L'Evolution Economique* (October 1913). In *MKNPR,* pp. 610-612.

57. Press announcement from the board of directors of the Russian General Oil Co. about the results of its activities for the last year and a half (issued in early 1914). In ibid., p. 624.

58. S. Pennaccio to A. G. Rafalovich (February 27/March 12, 1914) in ibid., p. 629.

59. Ibid., pp. 738-739; D'iakonova, pp. 131-33.

Chapter VIII

The War and Postwar Peace

The flames of the world war blazed for over four years and drew many countries into its destructive orbit. To some extent, military activities neutralized the oil competition or, more precisely, subordinated activities of corporations to the interests of the combatants. This meant a breathing spell in the oil conflicts, but the return of peace in 1918 ignited an unprecedented explosion of competition.

OIL CORPORATIONS DURING THE WAR

Not long before the first volleys resounded in August 1914, at a meeting in Berlin (in July 1914) the director of the Deutsche Bank, Gwinner, took aside the representative of Nobel (K. W. Hagelin) and told him: "We'll soon have war with Russia." Gwinner accused the tsarist government of not wishing to conclude a commercial treaty with Germany.[1] This episode serves to confirm how closely, on the eve of war, the activities of corporations, banks and industrial companies were interwoven with larger state politics. The petroleum companies and allied financial institutions were drawn into the preparations for war and had an impact on them. At the same time, they experienced the direct influence of political factors, which, from the outset of the war, acquired ever greater significance. The polarization of political forces that the war provoked made itself felt in the petroleum competition and subjected the oil business to the special conditions of wartime.

Under wartime emergency, the actions of oil corporations were subjected to strict regulation. At the same time, the preparations for war and then the armed conflict, as already noted, provided powerful stimuli to the growth of the oil business and the maturation of the modern form of monopolistic conglomerates. This should be especially emphasized, since the military factor continued to exert a major influence on the structure, character, and scale of activity of the corporations in later historical development.

One of the most important consequences of World War I was the increase in the scale of the oil industry, which proceeded with the active support of the government. Thus the Nobel firm had doubled its capital stock from 15 to 30 million rubles on the eve of the war, but by 1916 it had expanded again to 45 million rubles. Nobel strengthened its position on the Russian market and increased its ties to the government and continued to occupy a leading position in the Russian petroleum business. Alongside the Nobel firm were two other influential groups—the bloc of ex-Rothschild properties (since transferred to Deterding, with substantial expansion) and the Russian General Oil Corporation.

In England the successful activity of the Anglo-Persian Oil Company was guaranteed by an agreement in 1914 with the government. The Royal Dutch Shell, which accumulated gigantic means from its deliveries to the Admiralty, belonged to the same category. American companies likewise experienced strong growth during the war years. Insofar as the United States did not enter the war until April 1917, Standard Oil and other U.S. companies deemed themselves free to trade with both combatants, which permitted them to reap fantastic profits. But England soon established a blockade on its adversary's ports and thereby deprived the Americans of the opportunity for windfall profits from supplying oil to Germany.

The countries of the Entente proved to be in an incomparably superior position over the Austro-German bloc, since they were supplied with an abundance of petroleum. Not in vain did Lord Curzon declare that "the allies floated to victory on a wave of oil." At the outset of the war, only 30 percent of the British fleet had been converted to the use of petroleum fuel; by its conclusion, however, 90 percent of the British navy used oil for its fuel. Only at one point—under the blows of a merciless submarine war—did the oil supplies to England appear to be endangered. The worst fears of the Admiralty, expressed in the course of debates about the expediency of converting the fleet to oil, became reality.

As a result of attacks by German submarines, several tankers were sunk and England's communications with overseas petroleum sources were seriously impaired. That problem, however, was rather quickly solved by organizing convoys of oil tankers. In the final analysis, it was not the oil deliveries from Persia or the Far East, but what came from America that proved decisive for England's fuel balance. German submarine warfare severely affected the interests not only of England, but also of the United States. Indeed, it proved one of the direct causes for America's entry into the war. In March 1917 the American *Oil and Gas Journal,* reflecting the opinion of oil industrial circles, declared that American oil men were "not alarmed over the prospect of war" between the United States and Germany.[2] On April 6, 1917 the United States declared war on Germany.

The United States accelerated oil deliveries to the allies, increased the number of tankers involved in oil shipping, and provided them with armed protection. In September 1917 a representative of the British command observed: "Without the aid of oil fuel from America our modern oil burning fleet cannot keep the seas."[3] However, this concerned not only the fleet, but also land forces as well as air transport, which assumed an important place in the armed forces of the allies. Along with England, France also showed a keen interest in American oil deliveries, experiencing an acute shortage of fuel for its motorized divisions, which, by the end of the war, included more than 50,000 trucks alone.

By the end of the world war, motor vehicles had become the most important means of transportation on the land theatre of military action. The first steps in the development of motor-vehicle production were taken in the United States at the beginning of the twentieth century. In 1902 the United States already had 23,000 vehicles, and by the eve of the war their number had mushroomed to more than one million. At first, motor vehicle transport developed chiefly as automobiles, but with the onset of the war, the production of trucks increased sharply. Whereas only 23,500 of the vehicles produced in 1913 (were one-twentieth of total output) were trucks (the remainder being automobiles), by 1918 the truck output had risen to 227,300 vehicles (a quarter of the

943,400 vehicles manufactured that year). In 1915, for the first time, the consumption of gasoline exceeded that of kerosene.[4]

Motor vehicles substituted for the shortage of railroad cars and went to places where railway lines were nonexistent. In the fall of 1914, during the Battle of the Marne, the French command even used taxis to ferry troops to the battlefront. But the main role in the war fell to trucks; the war, without doubt, signalled the victory of the truck over the train. In August 1914 the English expeditionary forces counted a total of 827 motor vehicles and 15 motorcycles; by the end of the war they already possessed 79,000 motor vehicles and 34,000 motorcycles, as well as armored cars and tanks.[5] The first airplanes also took to the air.

England and France successfully organized their own production of motor vehicles, armored cars and everything else that was necessary for the mechanized units. And they had at their disposal the colossal arsenal of the United States. Above all, it was essential that all this new technology be provided with the requisite fuel.

The position of Germany and her allies was quite different. They also had ships with oil-burning engines, aviation, trucks and armored cars. But from the very outset they began to experience an acute shortage of fuel. Germany's annual consumption of oil before the war had been 1,250,000 tons. Of that volume, 77 percent came from the United States and 3 percent from Russia; from the outset of the war, none of that oil was any longer available. Only Rumania and Galicia, which had supplied the remaining 20 percent of Germany's prewar oil, were still helping to satisfy Germany's petroleum needs.[6]

Germany's crisis was so acute that it even attempted to obtain deliveries of Mexican oil. By inciting anti-American sentiments in Mexico, the Germans hoped to foment a military bloc against the United States. In the event they succeeded, they hoped to secure not only a bridgehead to conduct war against the United States, but also a source of supply for oil (by then Mexico accounted for about 7 percent of the world output). But the scheme, however tantalizing, was hardly feasible.

Equally adventuristic were the German designs on the islands of the Dutch Indies. Shortly before the outbreak of war rumors spread that Germany was negotiating with Holland for the purchase of Sumatra, whose oil fields provided about half of the entire output of the Dutch Indies (almost 1 million tons per annum). Having learned from "a source that merits attention," the tsarist Ministry of Foreign Affairs charged its ambassadors in The Hague, London and Berlin "to verify in a confidential manner the veracity of this information."[7] They were not able to clarify anything, however. Only many years later did archival materials of the German Kaiserreich confirm that the seizure of the Dutch colonies entered into the plans of German expansion and that, in the course of the world war, the Germans repeatedly came back to this question.[8] To be sure, the Netherlands was not a participant in World War I and remained a neutral throughout. But the oil fields of her colonies were in the hands of the Anglo-Dutch trust, Royal Dutch Shell, which delivered the oil to the British Admiralty.

In any event, Germany's main source of oil proved to be Rumania. Until August 1916 she remained a neutral bystander in the war. Indeed, in October 1914 the Entente countries even purchased a large quantity of gasoline from Rumania, and in November signed a contract with a Deterding branch (Astra-Romana) for future deliveries.[9] Nevertheless, the bulk of Rumanian oil went to the Central Powers; in just one year (from July 1914 to July 1915), Rumania delivered 613,000 tons of oil to Germany and its allies. In August 1916 Entente diplomats succeeded in drawing Rumania into military actions on their side, but the Rumanian army—poorly armed and poorly trained—began to retreat under the onslaught of the German troops. By the end of 1916 the Germans had occupied three-quarters of the Rumanian territory. Before retreating, however, the Rumanians destroyed and burned the oil-bearing regions. Special brigades under the supervision of English officers filled in more than 1500 oil wells, burned about a thousand wells and derricks, blew up reservoirs with 150,000 tons of oil and some 70 refineries, and put the torch to 830,000 tons of gasoline and kerosene. At the order of the head of the French military mission, pipelines and storage facilities in the port of Constance were also destroyed.[10]

Germany energetically set about restoring the Rumanian oil economy and in the next two years succeeded in extracting more than 1 million tons of oil.[11] But that was still too little; Germany and her allies suffered constant shortages of petroleum fuel for its naval vessels and mechanized land units.

To compensate for the shortage of oil products, Germany attempted to produce liquid fuel from brown coal through an improved method of coking. By the end of the war the annual production of this surrogate reached two million tons. It was used as liquid fuel for the navy, including submarines. In the fall of 1918, at the moment of Germany's capitulation, the reserves of this product amounted to six months' supply for the submarine fleet. Germany also began to produce a surrogate gasoline. All gasoline and its surrogates were reserved for aviation, but the supplies proved inadequate even for that limited purpose. Fuel was likewise inadequate to cover the demand for trucks and tanks. Lubricants were also in short supply. In brief, despite desperate attempts to reestablish production in Rumania and to produce a liquid fuel from brown coal, Germany experienced an acute shortage of fuel throughout the war. That shortage posed a serious impediment for the army and military operations.

The situation of the Entente countries was quite different. Only in early 1917, as a result of Germany's merciless submarine warfare on enemy shipping, did the supply of oil to the allied armies appear to be threatened. Shortly after the United States entered the war, an Inter-Allied Oil Committee was formed to regulate oil deliveries and, above all, the supply of liquid fuel for the British navy, whose monthly consumption amounted to 600,000 tons. England obtained approximately 80 percent of its oil from the United States.

The Inter-Allied Committee was also responsible for supplying gasoline for the allies' mechanized units on the European theatre of military operations, but it seemed to regard that task as a secondary priority. The result was irregular, inadequate deliveries of gasoline. "We must receive oil," said the commander-in-chief of the French army, General Foch, "or we will lose the war." In December 1917 the Prime Minister of France, Clemenceau, was forced to send a special communication to the American President and urge him to render the necessary assistance:

> At the decisive moment of the present war, when in the coming year of 1918 crucial military operations will commence on the French front, the French army should not for one moment be subjected to the risk of being left without gasoline. Every shortage of gasoline can suddenly paralyze our army and force us to conclude a peace unacceptable for the allies.

Clemenceau noted that the minimal reserve of gasoline, which should be at the disposition of the army (which consumed 30,000 tons per month), was set by the military command at 44,000 tons. But at present it actually had only 28,000 tons and this could "soon fall to zero." "If the allies do not want to lose the war," Clemenceau wrote in conclusion, "they should be concerned to see that a fighting France, at the decisive moment of the German onslaught, receive the gasoline which it needs, like blood, for the coming battles."[12] Wilson transmitted Clemenceau's request to the Committee on Petroleum, which was headed by one of the directors of Standard Oil, A. Bedford.

In 1915 the American deliveries of gasoline to Europe constituted slightly more than 0.5 million tons, but by 1917 this volume had doubled to more than 1 million tons. In 1918 the gasoline consumption of mechanized units constituted 100,000 to 140,000 tons per months, of which 90 percent came from the United States. Thus Bedford's organization provided a steady flow of petroleum products to the allied forces in Europe.[13]

The Committee on Petroleum had been established in the spring of 1917 by the Council of National Defense with the task of preparing for America's entry to the war. All countries interested in regulating the supply of oil created committees to control the delivery of this most important strategic product. Everywhere they were established at the initiative and with the direct participation of governments. In Germany, a single organization for oil supplies encompassed companies of various orientations and acted in accordance with wartime regulations and under the strict control of the government. In France, a committee of the ten largest oil refining enterprises was charged with providing supplies of gasoline and acted under the constant supervision of the government.

In Russia, a Special Conference on Fuel was created to control deliveries to branches of industry and transport that "work for goals of state defense."[14] As a result of the "oil famine" that existed even before the outbreak of war, the problem of providing liquid fuel for the fleet proved especially difficult. The United States Secret Service reported that Russia was forced to substitute coal bricks for oil on its naval ships.[15] The shortage of oil proved so acute that in September 1912 the Tsarist naval minister, I. K. Grigorovich, proposed a total ban on its export.[16] Like other powers, Russia paid serious attention to the petroleum needs of its navy. Even at the end of the nineteenth century, the Russian naval attaché in Washington (D. T. Mertvago) was corresponding about this subject with authorities in the U.S. Navy Department.[17] During the war, the Russian government established a "Special Conference on Fuel" (August, 1915), which was endowed with extraordinary powers. For all practical purposes, this was a government organization, comprised chiefly of representatives from different ministries and headed by the Minister of Trade and Industry. Thus, it also included representatives from monopolistic organizations, but the latter did not set the tone of work in the committee.

The situation in the United States was diametrically opposite to that in Russia. To be sure, Bedford's committee arose under similar circumstances, with the task of arranging oil deliveries under wartime conditions. However, the committee's leadership fell into the hands of Standard Oil Co. of New Jersey. After the death of John Archbold in 1916, Bedford became president of the company. It was precisely Bedford who, after resigning from his position as president of Standard Oil, became head of the Committee on Petroleum, which also included the Rockefeller directors Teagle, Lufkin, Moffett, and Doheny. The committee included as well representatives from companies outside the sphere of influence of Standard Oil—the Gulf Oil Corporation and the Sinclair Oil and Refining Co. But, for all practical purposes, the committee was run by Standard Oil and even made Broadway 26 its official headquarters. Bedford established direct contact with the British, French, and Italian embassies and played the role of a minister for oil deliveries. A few months after the committee's formation, attempts were made to reconsider its status. When the Council for Defense Affairs was replaced by a new government organization (Committee of War Industries), the latter attempted to absorb the Committee on Petroleum. But nothing came of its efforts. Relying on the support of industrial circles, Standard Oil appealed to the government and succeeded in defending its position. Admittedly, it had to make some concessions—to include in the committee (renamed the Committee on Fuel) representatives from the government, but even so the position of Standard Oil remained predominant.

The wartime committees on oil deliveries, established in the United States and other countries, had the aim of mobilizing resources and regulating both production and trade. Although the degree of state intervention varied from one country to the next, economic regulation appeared essential under the extraordinary conditions of wartime. The collaboration of corporations and governments also promoted a tendency toward fusion of the state apparatus with monopolies, a development that was exceedingly important in the subsequent evolution of state-monopoly capitalism. Over the next ten to fifteen years this pattern became widely disseminated and came to form an integral part of the system of contemporary capitalism. But the first impulse in this direction was a consequence of World War I.

The committees for oil deliveries were created for purposes of control and regulation, including prices. In practice, however, they created favorable conditions for the enrichment of oil corporations, which received colossal, unheard-of profits. This phenomenon was most evident in America where, in Lenin's words, as a result of World War I "they amassed more than everyone" and became "through their tributaries the richest people in the country."[18] Among those who earned billions of dollars were the American oil magnates, with Rockefeller's Standard Oil at the head. According to official data, pure profits of the 32 companies comprising Standard Oil increased severalfold and by 1918 had risen to 450 million dollars, of which a substantial part came from exports abroad.[19]

Although operations on the world market always occupied an important place in the strategy of the American oil trust, during World War I this business proved so profitable that 26 Broadway began to consider whether "would it not be worthwhile to divert our main attention to the international arena?" "It seems to me," said E. Sadler, one of the Rockefeller directors, "that the future of the Standard oil Company, particularly the Jersey Company, ... lies outside of the United States, rather than in it." Developing this idea, he continued: "If the policy of foreign expansion is admitted in principle, it seems to me that the Jersey Company is by all means the one indicated to handle it and that we should be efficiently organized."

Together with the enormous deliveries of oil to Europe, the firm also restructured its commercial operations in Latin America.

Rockefeller had made his very first attempts to establish a base there in the early twentieth century. But Standard Oil's share of operations in this part of the world was relatively small; suffice it to say that American oil exports to Latin America comprised just 5 percent of the output. As for the exploitation of oil reserves, Rockefeller participated even less. The sole country that, by the beginning of the century, had organized the industrial exploitation of oil was Peru. However, right until 1914, the oil company that worked these resources belonged to the English; it was only after the onset of World War I that Standard Oil acquired these properties. Venezuela has exceedingly rich oil fields, but these were not discovered until much later, remaining unknown until the 1920s. True, from the late nineteenth century Venezuela began to produce asphalt and attracted some interest on the part of American capital. In addition, soon after Royal Dutch Shell discovered oil reserves on the neighboring island of Trinidad (a British colony) in 1909 and began their exploitation, an exploratory company was created in Venezuela, with a leading position again occupied by Deterding.

On the eve of World War I, American investments in Latin America were far less than those of the British—1,242,000,000 compared to 4,983,000,000 dollars respectively.[21] Nevertheless, the United States firmly insisted that they have priority rights in Latin America. This concerned not only their relations with West European countries, but with other Latin American states. American diplomacy was not bashful about giving orders to Latin American governments and forcing them to accede to its demands. For example, in 1900 a local Venezuela company obtained a concession to produce asphalt, but insofar as the latter affected American oil interests, the United States Government threatened to use military force if the concession were not annulled.

The Russian ambassador in Washington (Kassini) reported that "at one time they preferred here even to send to Venezuelan waters an American squadron," but later confined themselves to "sending one naval vessel to the shores of Venezuela." This was a typical example of gun-boat diplomacy, and Kassini was correct when he wrote that this case reflects "the characteristic attitude of the federal government to such questions" and that it serves "as a clear demonstration of how zealously the American government protects everything involving the commercial interests of the United States and how little it stands on ceremony with the small South American republics."[22]

On the eve of World War I, the countries of Latin America became one of the most important markets for oil sales, the main suppliers for which were the Standard Oil Company of New Jersey and the Royal Dutch Shell Company. They built oil storage facilities in port cities and commercial centers and organized distribution networks for retail trade. They also constructed refineries, which became the outposts and supports for oil trusts in this part of the globe. Standard Oil built the first oil refinery in Latin America (on Cuba) in the late nineteenth century, and by the beginning of World War I Standard Oil had opened distilleries in the largest countries of South America—Brazil and Argentina. By this time Standard Oil had numerous branches and refueling stations in Latin American countries, and its income from trade here constituted 17 percent of the Company's total profits.[23]

By 1914 Standard Oil had established branches and agencies in Brazil, Cuba, Peru, the West Indies, Chili, Argentina, Bolivia, Ecuador, Venezuela, British Guiana, Columbia, Panama, Nicaragua, Guatemala, Paraguay, and Uruguay. That provided the basis for a steady expansion in its oil trade. In 1918, compared with the prewar period, the annual income of Standard Oil from operations in Latin America rose from 17 to 40 percent of the total profit that Standard Oil received abroad.[24]

Another important object of expansion for Standard Oil during the war years was the Far East, with its boundless markets and rich oil reserves. After many years of unsuccessful attempts, the Americans finally managed to acquire promising oil fields in the Dutch Indies. In general, the situation in this part of the world did not favor the plans of the United States. The "21 demands" that Japan had imposed on China in 1915 made it difficult for American capital to penetrate the Far East. In 1917 the United States obtained a commitment from Japan to observe the principle of open doors, and that inspired certain hopes. But the difficulties persisted, both because of Japanese policies and because of the opposition that American expansion encountered from the side of Royal Dutch Shell.

American Standard Oil and the Anglo-Dutch trust of Deterding combined their forces in organizing oil deliveries to the Entente.

But clashes between them, even under wartime conditions, persisted. In the fall of 1917 the conflict between Standard Oil and Royal Dutch Shell became so intense that official representatives of England and the United States had to intervene and mediate the dispute. The cause of the conflict was the increased deliveries of American oil, which Deterding sought to prevent. The issue was resolved in favor of Standard Oil, whose services were absolutely essential to England.[25] In addition, the affair occurred amidst a campaign in England against Royal Dutch Shell that had been provoked by representatives of the Foreign Office, which had ties to the Anglo-Persian Company. In March 1916, in a speech to the English trade congress, the head of the Anglo-Persian Co. (C. Greenway) declared: "It is my conviction that for patriotic reasons it is necessary to check the further growth of this vast foreign monopoly [Royal Dutch Shell], inasmuch as it has become a serious national menace [to England]." The leaders of the Anglo-Dutch firm were blamed for the fact that its branches in Rumania and in Scandinavian countries were delivering petroleum fuel to Germany and her allies. Particular enmity was expressed by the Foreign Office and its division responsible for the struggle against contraband. Shell's representative in London was forbidden to appear in the buildings of the ministry of foreign affairs, and could address the government only in writing.[26]

The accusations against Royal Dutch Shell, however, were tendentious. In March 1915 Deterding left the post of chairman in Astra-Romana, declaring that the firm "is providing the enemies with contraband." In 1916 British military intelligence devised plans to destroy the Rumanian oil works, and it relied upon the direct participation of Shell's representatives. The army and navy were both interested in cooperating with the Anglo-Dutch trust. Its enterprises in the Dutch Indies produced toluene, which was then used to manufacture explosives (trinitrotoluene—TNT), and produced the gasoline that was needed for trucks and planes. The Admiralty's representatives described Royal Dutch Shell as "excellent contractors whose usefulness we appreciate," as "most efficient and indispensable contractors," which "invariably rendered every assistance and have never attempted to erode the conditions of any contract which has been entered into."[27]

Although wartime conditions, to some degree, narrowed the differences between the two oil corporations, the endless squabbling and fighting between them did not cease. The same was true of the allies: despite the wartime solidarity, the struggle over the oil question persisted. On the eve of war, the French ambassador in London (Cambon) came to the conclusion that the British policy is aimed at seizing the world's oil deposits.[28] France gave the English to understand that she would not remain indifferent to this question. Negotiations were initiated that ultimately led to the Anglo-French agreement of Sykes-Picot (May 1916) dividing spheres of influence in the Near East. That agreement laid the foundations for new agreements on the oil question, whose term did not expire until after the end of the world war.[29]

A still more serious conflict erupted between England and the United States. Although their differences were muted by wartime conditions (and, above all, by the mutual interest in American oil deliveries), the grounds for subsequent conflict steadily accumulated.

British higher circles looked upon the seizure of oil reserves as one of the prices of the world war. The logic of British policy was thus: the more London became convinced of its dependence upon American oil, the more intensely it sought alternative sources.

To assert its hegemony in the Near East, before the war England had already instigated uprisings by Arab peoples against the Turkish empire—which had been in constant conflict. A widespread network of British agents was active during the war in the Arab East, and one of their primary goals was to advance British oil interests. An example is provided by the well-known role of Col. Lawrence, who succeeded in winning the confidence of Arab tribes and coordinated their military activities. During the war military activities also spread across the Near East and involved not only regular troops but also Arab rebels. The seizure of oil-bearing regions and establishment of British control was the most important objective of England's military operations in the area. After the defeat of the Turks and the occupation of Palestine in 1918, the English command forbade American activities there.

However, one month before America's entry into the war, Standard Oil's board of directors worked out a program of action to expand the American oil holdings. It consisted of these points:

> We should know what is going on in all foreign fields as to development work and transfers of property and we should be before the eyes of all owners of property who might sell or lease.... We should collect data as to the future possible producing areas in the world and interest ourselves in the most promising.

This program was presented in a memorandum, which was discussed by the board of directors and given to management as a guideline.[30]

The success of the allied armies and the role that American oil deliveries played, and also their financial and economic support in securing the Entente's victory, encouraged the growth of nationalistic passions in the United States and engendered confidence in America's right to a special role in world affairs. To a prosperous America it seemed that Europe was old and feeble, a superannuated creature that had outlived its time. Shortly after the end of the war, Standard Oil sent Secretary of State Lansing a confidential memorandum. It declared that:

> Europe generally is still floundering in the morass of after-war conditions. There is no tangible evidence that any of the countries have raised themselves in any appreciable degree, while on the other hand, it is undeniable that certain countries have sunk deeper.

The memorandum drew the conclusion that, under these circumstances, Europe "wants to listen to American advice" and "if we can get our businessmen really interested in these problems, they can find a solution."[31]

The American government endorsed his call. To American representatives abroad, it sent the following instruction:

> Gentlemen!
>
> The vital importance of securing adequate supplies of mineral oil both for present and future needs of the United States has been forcibly brought to the attention of the Department. The development of proven fields and search for new areas is being aggressively conducted in many parts of the world by representatives of different countries and concessions for mineral oil rights are being actively sought. It is necessary to have the most full and fresh information regarding the activity of both United States citizens and others. You are accordingly instructed to obtain and forward promptly information regarding mineral oil concessions, change of ownership of oil

> property or important changes in ownership or control of corporate companies concerned with oil production or distribution. Information regarding the development of new oil fields or increased output of producing areas should also be forwarded. Comprehensive data are desired and reports should not be limited to the points specifically mentioned above, but should include information concerning all interesting matters pertinent to the oil industry which may arise from time to time.
>
> You are also instructed to lend all legitimate aid to reliable and responsible U.S. citizens and interests who are seeking oil concessions or rights. Care should be taken, however, to distinguish between U.S. citizens representing American capital and U.S. citizens representing foreign capital; also between companies incorporated in the United States and those companies which are merely incorporated under U.S. laws but dominated by foreign capital.[32]

Official American representatives abroad had always actively supported oil interests, but the above instruction from the State Department gave a new fillip to this long-standing tendency.

From Vladivostok and Tientsin, Capetown and Jerusalem, Madras and Tunis, Sidney and Algiers, Maracaibo and Veracruz, La Paz and Lima—came telegrams, dispatches, and reports. A stream of correspondence flowed in from the Middle East. To all the more important points of this region the State Department sent a supplementary instruction: "Cable or regularly write about all events, petroleum rights, options, prospecting, etc. as per your advice as to action for safeguarding American petroleum interests vested or potentially, especially desired."[33]

It turned out that the English vigilantly followed every step the Americans took and wherever possible sought to hinder them. It seemed to some that England had lost its former strength. In a confidential memorandum to the State Department, Standard Oil asserted that in certain respects the situation in England is "worse than anywhere else."[34] Such evaluations, however, substituted fantasies for reality.

The English emerged as the main rival of Rockefeller. For several years, complained a representative of the American geological service, Great Britain has gradually been preparing to seize control over the world's oil resources. Two board members at Standard Oil, Teagle and Bedford, confirmed the "mounting tension." In their words: "On the West flank of the series of mountain ranges which extend almost uninterruptedly in a northwest-southwest direction from Smirna in Asia Minor to the mouth of the Ind in India, there are prospective oil areas which are believed to be of immense potential value as regards both the quantity and quality of production."[35]

The problem was that the Americans were denied access to these oil deposits. In mid-May 1920 the American ambassador to London, D. Davis, and the representative of Standard oil, L. Thomas, met with J. Cadmen from the British Oil Commission. When Davis inquired "if his statement was tantamount to an open door policy," Cadmen "did not reply directly, but fell back on the statement that governmental policy was not settled and he could not be hurried in the matter." However, a little while later the State Department learned that the words of the British representative were a pretext for evading the question.[36]

In 1918 the Americans produced 65 percent of the oil output of capitalist countries. According to the data of the American Geological Service, American oil reserves were on the verge of exhaustion, while the domestic demand for oil had virtually doubled during World War I. Therefore America grew anxious about the future prospects of satisfying their own demand for petroleum products. In 1920 the price on crude oil increased by 50 percent and stood three times higher than the 1913 price level. J. Smith, the director of the U.S. Geological Survey, saw only two possible solutions: either reduce oil consumption in the United States or depend on foreign sources.[37] It was difficult to expect America to reduce its oil consumption. Even during years of war, when, because of the need to increase deliveries to the Entente, it was proposed to limit the use of fuel by private automobile owners, the most that could be achieved was to forbid the sale of gasoline on Sundays. By the time that the war had ended, the number of automobiles and trucks in the United States had reached 6 million, rose to 10 million two years later, and climbed to 15 million after another three years (1923).[38] The consumption of oil products, correspondingly, increased significantly.

There thus remained the second solution—to seek oil abroad. And it was precisely in this direction that American oil corporations, with Standard Oil in the avant-garde, made their main efforts, embarking on a bitter struggle with their rivals, above all the British corporations.

THE INCREASING RIVALRY IN THE POSTWAR YEARS

As soon as peace reigned and the canons fell silent on the field of battle, the war for oil immediately began anew. Its arena, as before, was the entire world. Along with England and the United States, France also participated actively in this struggle. Insofar as oil corporations worked hand-in-glove with official diplomacy, this imparted a particular seriousness to the conflict. In contrast to the previous period, when the struggle had focused primarily on the control of world petroleum markets, the central question now was that of seizing and dividing up petroleum reserves. The main locus of conflict was the Near East, where the rivalry was interwoven with the question of the Turkish legacy—that is, the political settlement for territories formerly part of the Ottoman Empire. After World War I, these areas wcrc divided into nominally independent Arab states under the mandate of the League of Nations.

To strengthen its claims to the concessions of the Turkish Petroleum Co. (acquired on the eve of the world war), England joined forces with France. The concession was located on the territory of Mesopotamia, which had been cut off from Turkey as a result of the war. At issue was the German share of the Turkish Petroleum Company, which amounted to 25 percent of its stock. To avoid penetration by the United States, England offered the German stock to the French. Responsibility for implementing this plan was entrusted to Gulbenkian and Deterding, who entered into negotiations with French financiers as the Banque de l'Union Parisienne (which had been formed with the participation of Royal Dutch Shell).

In a secret meeting at San Remo in April 1920, France and England reached an agreement to resolve their oil disputes. England, which had obtained a mandate from the League of Nations as overseer for Arab states in this region, established itself in Mesopotamia (soon renamed Iraq); France's sphere of influence was Syria, through whose territory France received the right to run an oil pipeline from the oil fields of the Turkish Oil Company. And, finally, France received the German share of stock.[39]

The goal of this agreement was to divide out spheres of influence in the Near East and, simultaneously, to exclude the United States from the region. After obtaining the text of the Anglo-French agreement, Standard Oil sent it to the State Department with a request to protest the actions of the European powers.[40] At a meeting convoked afterwards, the representatives of the company demanded "assurance of a stronger foreign policy on the part of the government."[41]

The government responded that it was already taking measures, that a protest note had already been sent to the British government announcing the United States' refusal to recognize the agreement at San Remo. It also accused England of violating the principle of equality and insisted that all who won the world war have an equal right to share the spoils of victory. The British Foreign Office was forced to agree with this point of view, but insisted on the special rights of the Turkish Oil Company, since it had already received this concession before the war. The U.S. State Department rejected this explanation: the English received the concession from the Grand Vizier, whose actions were illegal, and it was obligatory to observe the principle of equality.[42] Standard Oil welcomed the "firm attitude on the oil question," and emphasized that "in the interests of all, of England, America and the Allies, the only sane and safe policy was that of the open door."[43]

The British government had to maneuver in response. In conversations with diplomats and representatives of Standard Oil, the English declared that their actions had been misunderstood. A special envoy sent to Europe after San Remo submitted a report to the government. He reported that in both London and Paris not only government authorities but also "large financial interests" are "most anxious" that the U.S. government take the initiative "by suggesting some basis of cooperation and understanding."[44] For both financial-economic and political reasons, England did not wish a dispute with the United States. At a meeting of the International Commercial Congress in London in the summer of 1921, Cadmen—in the presence of a representative of Standard Oil—declared that he was disappointed by the sharp tone of the diplomatic notes. He proposed "an amalgamation of American and English private interests in the development of the oil fields in the Middle East."[45]

This proposal was difficult to realize, for England did not intend to make serious concessions and the Americans were not placated with symbolic gestures of conciliation. On the territories separated from the former Ottoman empire, numerous British agents remained active and in constant touch with Arab tribes. "Arab politics is extremely complicated," reported A. Dulles from Constantinople during this time. A bitter struggle was being fought among the different tribes and groups. Although from the outside they seemed to reflect national differences, in fact (as Dulles noted) "here oil rather than ethnic considerations will influence both parties."[46] In the Arabian Desert the English intelligence agent, Col. Lawrence, remained active; in Constantinople, in the chancellery of the American Commission for the Near East, was A. Dulles, a future super-spy who was only beginning his career. As he wrote, "I have only begun to try to do some reading to straighten out the various currents and forces which are behind the present happenings in Arabia."[47]

The United States endeavored, at any cost, to attain success. The State Department made a special study of this question and came to this conclusion: "The formula 'open doors' does not aid very materially, because in most countries the supply of petroleum is so limited that the door cannot remain open except until two or three companies have taken control of the resources. What we need ... is to regard the open door as merely prefatory and to proceed toward a definite position on the petroleum question in each region."[48]

This question was widely discussed in the business and political circles of the United States. Various solutions were proposed, all aimed at creating a counterpoise to English expansion and to promote American interests. In May 1920 an attempt was made to organize a government oil corporation in the United States after the model of the Anglo-Persian Oil Co. Although that attempt did succeed, a year and half later—in accordance with a proposal from the then Secretary of Commerce, H. Hoover—a syndicate of the seven largest American oil firms was created: Standard Oil of New Jersey, Standard Oil of New York, Texas Oil, Gulf Oil, Mexican Eagle, Sinclair Oil, and Atlantic Oil. This syndicate was supposed to defend American interests in the Near East. Very soon, its representatives began negotiations with the Anglo-French group and an agreement was reached for admission of Americans to the Turkish Petroleum Co. In addition, the share of stock assigned to the American syndicate was allocated from the share of the Anglo-Persian Oil Co. All the largest oil companies participated in this agreement, and it can justly be regarded as the foundation of "a long-term plan" to acquire the oil reserves of the Near East. The Iraq Petroleum Co., formed quickly on the basis of the Turkish Petroleum Co., truly "set the pattern for the Middle East concessions of today."[49]

While the Near East was the main theatre of oil wars in the post-war period, the second most sensitive hot spot was Latin America. As before, attention was concentrated primarily on Mexico. And here the conflict with England constituted an important part of the problem. No less important was the role played by domestic Mexican affairs—the further development of the Mexican Revolution. Carranza, who had replaced Huerta, continued to govern the country. Not long before he came to power, a representative of American oil interests (N. Rhodes) wrote the Secretary of State, J.R. Garfield, that he "would feel rather confident of the success of any form of government in Mexico if actually backed up by the armed forces of the United States or combined powers." At the same time, he believed that "their own internal differences are now very serious and becoming more so each day, to such an extent that I have no confidence whatever in their ability to establish or maintain a stable government."[50] The American representative had in mind here a government that would suit the interests of the oil corporations. However, Carranza did not give the United States cause for such jubilant hopes. Although oil companies endeavored to establish contact with him, and the U.S. Government gave de facto recognition, the situation remained unstable.

Responding to direct pressure from oil corporations, certain circles of the United States continued to entertain the idea of intervening and overthrowing the Carranza government. "It is of course to be expected," as Garfield wrote Lansing, "that the representatives of great financial interests, which in bygone days dominated the councils of Mexico, will oppose the new government and possibly offer to form a new government based upon wealth and business interests." Yet, that prospect, however alluring it might have been, had no realistic basis. To quote Garfield: "Such offers mean nothing in the long run, because should such a government be established, it would within a short time face revolution. It is impossible to kill the new life in Mexico by a return to a system of government based upon the dominance of the wealthy class."[51] The situation was aggravated by the steady growth of anti-imperialist sentiments in the country. "Everywhere," reported Rhodes, "I saw bitterness against Americans as a class, but not against individuals."[52]

The growth of the national liberation mood led to the inclusion of an article in the 1917 Mexican constitution that aimed to protect the country's natural resources. The article declared that the land, mineral resources and water within the borders of the Mexican territory belong to the nation. The adoption of this constitution provoked new pressure on Mexico. Carranza's government fell, and A. Obregon became president. Attempts were made to reach an agreement with him, but without success. Not until 1928 was a compromise finally reached: according to a newly adopted law, the oil companies retained their property.

The events in Mexico exacerbated the tensions in Anglo-American relations. In December 1920 the London *Daily Telegraph,* referring to "information coming from the State Department," reported that the Americans were incensed by England's attempt to control the petroleum industry in Mexico. During these same days the *New York Herald-Tribune* announced that the dispute over Mexican oil "will lead to war between England and the United States."[53] In 1923 Deterding purchased the Pearson company, while almost simultaneously Doheny began to collaborate with the Standard Oil Company of New Jersey.[54] The Anglo-American conflict had entered a new, and more dangerous, phase.

The other Latin American country that provoked tensions and rivalry between the two countries was Venezuela. In contrast to Mexico, a reactionary dictatorship had firmly established itself here. The Venezuelan dictator J. Gomez expressed a willingness to meet the United States halfway. Through the American envoy in Caracas, he indicated that Standard Oil would be given preferential treatment. A company representative immediately paid Gomez a visit and received a cordial reception. True, when Standard Oil filed its official application for oil tracts, it was informed that these had been leased that very day to Gomez's son-in-law. However, this problem was easily rectified; the president's family imply wanted to receive a certain additional sum. Standard Oil made the obligatory pay-off. The tracts thus acquired provided the basis of "Standard Oil Co. of Venezuela," later renamed "Creol

Petroleum Co.," which, along with Arabian American Oil Co., became the Standard Oil Company of New Jersey's most important oil possession abroad.[55]

At the moment when the United States exhibited an interest in Venezuelan oil deposits (which, as soon became clear, were among the richest in the world), Deterding had already established a firm base in the country. Here too he emerged as Rockefeller's main rival. But the Gomez government kept its word and continued to give preference to the Americans on the grounds of the need to "balance" the British influence.

More militant American representatives declared that "if Venezuela should persist in disadvantageous conditions on American companies in this country, the United States government should force Venezuela, if necessary, to remove such conditions."[56] These words belonged to M. L. Requa, former General Director of the Oil Division of the United States Fuel Administration and later vice-president of the Sinclair Oil and Refining Company, who was "something of a crank on this subject" (according to the State Department), although he was well informed and had "no antipathy to any country except Mexico."[57] However, there was no need to resort to force. As the American historians G. S. Gibb and E. H. Knowlton have pointed out, the relations between American oil business and the Gomez government were "satisfactory." In Venezuela, in fact, "the oil industry enjoyed a sense of security unprecedented in degree in any other Latin American country.. .."[58]

These guarantees were based on the appropriate articles of the Venezuelan constitution. In addition, they were constantly buttressed by diplomatic pressure from the United States. This fact was conceded in an internal memorandum of the State Department, which noted that in the 1920s in some cases, "especially in Venezuela, the [United States] Government unofficially assisted American oil companies in receiving and retaining rights under concessions," but "this assistance does not seem to have given rise to any formal correspondence with the concerned governments."[59] Ten years later, the American envoy in Caracas wrote that "local authorities have absolutely no doubt as to the impartiality of the American government and the Embassy to all oil matters." He said that he "scrupulously avoided any show of favoritism" and that he had met "all the heads of the oil companies and each of them was in the Embassy." True, in this same dispatch the ambassador reported that the largest companies (such as the Standard Oil Company of New Jersey and Sinclair) "know local officials more or less intimately and do not ask the Embassy for anything."[60] As for the "impartiality" of the diplomatic service with respect to oil companies, it turns out that this referred more to form than substance. But such a situation was hardly abnormal; the interests of big capital always stood at the center of attention. No matter what the subterfuges and qualifications, favoritism remained the prerogative of big business. It was given preferential treatment, and its demands were defended by all possible forms of state intervention.

At the very beginning of the 1920s, this question was raised within the context of general national goals by the former Secretary of State Robert Lansing, who had left his post in 1920 and formed part of the political opposition to the Republican administration. He was the first head of the foreign policy organs of the United States who placed the goal of oil expansion within the framework of state affairs. "Up to recent years, the important questions in international affairs were political in nature," he wrote in a memorandum of American foreign policy, but then added: "It is not so today. Trade and business interests in foreign fields are now absorbing more thoughts than the political interests, which have so long controlled foreign policies." In the opinion of the former Secretary, the State Department "was slow to recognize this change.... American business seeking markets and foreign investments complained that the Department failed to aid it and to give it proper protection." Lansing demanded "practical and aggressive policies which will materially advance American interests and protect American rights. If the Department of State cannot change its character sufficiently to meet these desires, the nation will find some other agency to execute its will."[61]

But it is fair to say that there was no special need for Lansing's strictures and threats. At the very height of the oil rivalry in the 1920s, the Mexican newspaper *El Universal* published an article called "the Oil Minister," which was devoted to Lansing's successor as Secretary of State, C. E. Hughes, whose name became closely linked with the American policy of "oil imperialism." As *El Universal* wrote:

> Oil has come to be then a black ruler, to which arbiter [sic], little by little, the peace and war of the peoples have been subjected, and the oilmen have come to be the most powerful men in the world. Not only do they intervene and influence the relations between the states, but further, they are transcendental factors in the internal policy of the nations.

Although the newspaper's statement referred primarily to the relations between Mexico and the United States, its conclusions applied just as well to the entire policy of the United States. In a similar vein the American liberal journal *The Nation* declared that Hughes' policy was completely subordinated to "oil and the almighty dollar."[62]

Significantly, during these years, when the United States Government consistently began to support the oil corporations, an attempt was made to discard the restrictive anti-trust legislation. During the war years the power of corporations and their direct influence on government affairs had increased. Thus, when Bedford became chief of the government's Committee on Petroleum, a top official in Standard Oil assumed control over one of the most important domains of wartime America. Although Bedford did resign from his post as president of Standard Oil, he remained as one of the company's directors.

Soon after the war, his Committee on Fuel sent a note to the Justice Department protesting the existing anti-trust legislation. The Committee accused the government of taking actions that weakened the Standard Oil group and impaired its capacity to fight Royal Dutch Shell. "The American oil companies," declared the Committee, "should be encouraged and assisted by sympathetic Government cooperation in acquiring throughout the world possible oil territories to secure necessary supplies for the growing demand for oil products in the United States." The anti-trust legislation was denounced as pernicious and contrary to national interests: "The size of the Standard Oil group, capitalization or similar subject are absolutely extraneous and have no bearing upon this discussion." Moreover, the initiators of this address contended that, precisely because of the centralization of the Rockefeller company and its gigantic potential, it was necessary to direct the firm toward the conquest of new oil deposits. "The size and power of the Standard Oil Co.," declared the Committee on Fuel, "are such as to enable them to more than hold their own in future competition with foreign companies." It concluded that "if the Standard Oil companies were not hampered and handicapped by government restrictions, this would be undoubtedly true; whether it is true under existing conditions may well be open to doubt."[63]

Despite the obvious factual distortions and the aggressive tone of the note, it succeeded in persuading the government to disregard the anti-trust restrictions on corporate activities outside the country. As Secretary of State, Hughes requested his branch's juridical section to investigate the possibility of abolishing the antitrust legislation. He was told that it is quite possible, with respect to corporate activities abroad, not to apply these laws at all. "There does not appear to be any United States law," wrote the counsel of the juridical section, "prohibiting American oil companies from pooling their interests with the aim of exploiting foreign oil fields." In elaborating on this view in a special memorandum, he noted the importance of cooperation by American interests abroad, but under the condition that such associations "should always be controlled by American citizens." "It is believed," concluded the counsel, "that the oilmen should understand from the very beginning that if diplomatic support of the United States Government is to be expected, any corporation which they form must remain under the control of American citizens and cannot be the instrument of advancing alien interests to the detriment of Americans." When the President of the United States was informed of this opinion, he said that "pooling would be illegal," but nevertheless admitted that the government "only encouraged American oil companies to work together."[64]

PLANS FOR A UNITED FRONT AGAINST SOVIET RUSSIA

An important component of the world struggle for oil in the postwar period was the effort of large corporations to forge a united front against Soviet Russia. Following the October Revolution in 1917, the new Soviet government nationalized the Russian petroleum industry, thereby putting an end to the power of foreign capital. The former owners did not wish to accept the new state of affairs and strove to regain their "rights." Although their efforts proved in vain, they did not easily accept the new state of affairs or understand that the changes brought by the revolution were irreversible. It was thus no accident that the former owners of oil property joined the ranks of the most ardent foes of the Soviet state. They supported military intervention and, when this failed, instigated various kinds of anti-Soviet campaigns. It is not surprising that Deterding, who had possessed an enormous oil enterprise in prerevolutionary Russia, was one of the most zealous enemies of the USSR; indeed his hostility toward the Soviet government became particularly intense after his marriage in 1924 to Lydia Koubeyaroff, a White Russian emigrée. Although Standard Oil had not owned any property in Russia before the revolution, in 1920—heedless of the Soviet decree on nationalization and with the sanction of the State Department—the company decided to purchase the securities of Russia's large prerevolutionary oil company, that of the Nobel Brothers.[66] When the leaders of the American trust weighed the pros and cons of acquiring the Nobel firm, they recognized a certain element of risk. But in the end they chose to disregard the dangers.

There were several reasons they did so. First, they believed that Russia's oil resources were so important and great that the risk was worth taking. Second, they did not believe that the Soviet government could survive long or that it could cope with the task of governing such enormous country as Russia. When Adler visited Novorossiisk, Batumi, and Baku at the end of 1919 on behalf of Standard Oil's board of directors, he found that the political situation in Soviet Russia was unpredictable, yet the leadership at Broadway 26 was firmly convinced that the Soviet government would not last for long.[67] In early 1922 the board sent the State Department a letter, which, referring to information obtained from Nobel, asserted that the Bolsheviks' power was eroding and that "all indications from Russia are to the effect that the Soviets, to maintain their position, must absolutely surrender Communistic ideas and return private property...." This was sheer invention, the substitution of wishful thinking for reality. The American historians Gibb and Knowlton note that the "Jersey men, like a great many others, awaited the collapse of the Soviet regime with confidence and took steps to place themselves in an advantageous position when the collapse should come."[69] Third, when Standard Oil decided to purchase Nobel's enterprises, not the least important motive was its fear that delays could work to the advantage of England. If the English—Standard Oil's arch-rival in the struggle for oil—obtained the Nobel firm, they would have a virtual monopoly over the ideal business in Russia.[70]

For all those reasons Standard Oil finally decided to sign the agreement with Nobel, thereby acquiring a bloc of stocks with a face value of 11.5 million dollars in exchange for a payment of 9 million dollars. When Deterding learned of this deal, he immediately bought up the stock of several other Russian petroleum companies—those of Mantashoff, Lianozoff, Gukasoff, and Tsaturoff. By adding these companies to the possessions of Royal Dutch Shell, the head of the Anglo-Dutch trust sought to bolster his claims to Russian oil. In any event, that is how his actions were viewed by the Americans, who jealously observed their rival's behavior. The leaders of Standard Oil even began to suspect that thc British oil industrialists were prepared to negotiate with Soviet authorities for a separate agreement behind the Americans' back.

These suspicions grew particularly strong on the eve of the Genoa Conference (April 10 to May 19, 1922), which discussed the economic relations between the Soviet state and Western countries. Both at the conference's sessions and behind the scenes, the subject of oil acquired exceptional importance. On the eve of the opening sessions at Genoa, the American press published reports that Deterding had concluded an agreement with the Soviet government for oil concessions. The leaders of Standard Oil appealed to the State Department for assistance, and Secretary of State C. E. Hughes telegraphed the American ambassador in Italy (Child, who served as the American observer at the Genoa meeting) an order to verify the rumors immediately, and to report the results of his investigation.[71] Although the rumors were not confirmed (being denied by both the Soviet delegation and Deterding), they continued to circulate during and after the Genoa conference, causing concern not only for the American but also the French and other representatives.[72]

As early as December 1920, a memorandum from the Standard Oil's board of directors to the State Department (entitled "The Russian Situation") noted that England is playing a double game and seeks, by means of separate negotiations, to obtain advantages for herself. Notwithstanding the extreme character of the Soviet regime, charged the memorandum, the British government and business "probably are inclined to take advantage of whatever opportunities for the development of industry and trade that the present situation in Russia may offer." In this connection, Standard Oil criticized the position of the State Department, which in its opinion was not sufficiently flexible and did not take account of the changes in the situation.[73] At a special conference organized by the State Department, Standard Oil's representatives complained that the American government was responding too slowly to the changes that were occurring.[74]

Soviet diplomats skillfully exploited the differences among the participants of the negotiations. A special role in this respect fell to L. B. Krassin [Krasin], who, prior to his appearance in Genoa as a member of the Soviet delegation, had for a long time carried on negotiations in London.[75] Shortly before the Genoa Conference he gave the American representatives to understand that their hopes for a restoration of property rights were groundless, although he did admit the possibility of investment projects and expressed a willingness to grant the Americans even certain advantages in the negotiations.

Just prior to the conference, Standard Oil sent a memorandum to the State Department, expressing its confidence that the United States Government will take steps to create a united anti-Soviet front with England and France. It also passed on news from Nobel that the Soviet government would soon be forced to abandon its economic program, that Nobel had met with Deterding, and that they had agreed upon a common policy—to seek from the Soviet government a categorical recognition of the rights of the former property-owners.

The nationalization of industry in Soviet Russia served as "a dangerous precedent" and later could have an impact on American interests "in other countries" as well. Bedford used these very phrases in a memorandum sent to Hughes a few days after the conference in Genoa opened. He expressed his confidence that neither at the conference nor in the lobbies will anyone make attempts to reach an agreement with Soviet delegates for oil concessions.[76] Deterding sent analogous appeals to the British Foreign Office.

The American government demanded that the European powers boycott negotiations with Soviet representatives. However, the

Americans themselves, whether they wanted to or not, were gradually drawn into contacts and negotiations with the representatives of Soviet Russia. In the long run, business contacts would play a role in leading the United States to extend diplomatic recognition to the USSR, but that was still a decade off in the future. Standard Oil was one of the leaders of the Anti-Soviet policy in Genoa, but her own representative made contact with Krassin, met with him, and even conducted certain negotiations. The head of Standard Oil's Italian branch (Mowinkel) invited Krassin to his villa and held informal talks on the question of possible Soviet-American negotiations and agreements.[77]

The meeting in Genoa was a continuation of contact that had been begun as early as 1920, when the representative of Standard Oil visited Krassin in London, where he was currently negotiating with the English. The Soviet diplomat assessed this approach as an attempt by the Americans to "play a dirty trick" on their English rivals.[78] In early 1922 Bedford himself met with Krassin. Bedford declared that his side followed with interest every event in Russia that might attest to the reestablishment of the rights of private property.[79] However, the hopes of Standard Oil's representatives became manifestly delusive, encouraging some to call for a reconsideration of the company's policy toward the Soviets. It is in this connection, apparently, that the meeting of Mowinkel with Krassin at Genoa should be seen, for the American representative asked the Soviet delegate not to conduct any kind of negotiations with the English until the possibility of a Soviet-American agreement had been discussed. Mowinkel expressed the wish that Krassin come to New York for an exchange of views.[80]

Walter Teagle, by then the leading figure on Standard Oil's board of directors, also met with Krassin. Teagle was confident as he entered the conversation, for he thought that his business experience would make it easy for him to get the better of the Soviet "commissar." In recounting later his meeting with Krassin (in a discussion with his biographer, G. Gibb), Teagle said that "the Commissar apparently was somewhat unpolished, dressed in a quite incongruous manner, ... a far cry from the polished European oil diplomats like Gulbenkian, Deterding, Riedemann and the rest!"[81]

Of course, compared with the clothing of oil magnates (C. Gulbenkian, H. Riedemann, H. Deterding and Teagle himself), the suit of the Soviet diplomat was more modest. But those who knew Krassin emphasize that he always comported himself with ease and grace, and that he felt himself equally at ease at business negotiations and at formal receptions, where he was elegantly dressed. Here is the recollection of A. N. Ehrlich, who participated in the Genoa conference: "Before us stood a tall, well-proportioned man dressed in the European fashion, with a dark complexion and slightly grayish-white face. His energetic movements and speech somehow had an immediate impact on those around him. A small, trimmed wedge-shaped bear and his graying hair gave him, if you will, the appearance of a businessman."[82] Indeed, during his years in emigration, Krassin held posts in industrial companies abroad, had excellent practice as a manager in the large German corporation Siemens Schuckert, and now made effective use of this training. He had enormous experience as an entrepreneur and political activist and revolutionary.

However Teagle may have referred to him, those who met Krassin noted the brilliant knowledge and extraordinary abilities of the Soviet representative, who found a way out of any complicated situation and who possessed a rare sensitivity and quick reactions. Teagle was punished for his self-confidence and snobbishness. "The unpolished Russians," by Teagle's own admission, "held all the trump cards and played them with consummate skill. The Americans found the negotiations extremely difficult and frustrating." Neither Teagle nor the other representatives of Standard Oil succeeded in getting the better of the Soviet "commissar," as they had wanted. "These polished gentlemen, completely at ease at the courts and high places of two continents, made private ridicule of the Soviet Oil commissar..., his manners and his appearance. This was certainly one of the ironies of history because ... Krassin pretty much made fools out of the Western diplomats who were trying to get a foothold in Russian oil-producing territories."[83]

In the course of his work on the Teagle biography, Gibb came to the conclusion that the refusal to recognize Soviet Russia was "an enormous blunder" in the life of the oil magnate. Teagle and his partners at Standard Oil "considered the Russian revolutionists to be irrational, dangerous, unintelligent in business ways and doomed to early destruction." Hence Teagle "played a waiting game, always expecting that the communist movement in Russia would collapse and that a conservative government would return to Standard Oil its vast investments in Russia." But that of course was a miscalculation, and "Teagle in his last years admitted that this had been one of his greatest blunders." While Gibb agrees with that assessment, he points out that it is "a bit unfair to Teagle," who was obviously "receiving bad advice from his foreign experts and from his friends in Washington."

It bears noting too that Teagle "regularly sought the advice of the Rockefellers." Although "this fact was always concealed and even today ... there is a reticence to admit that this was the case," in fact the Rockefellers "were indeed consulted on major company moves, and certain company policies (such as a major change in employee relations) were initiated at the insistence of the Rockefellers."[84] Indeed, after J. D. Rockefeller formally retired and surrendered control to such managers as Archbold, Bedford, and Teagle, it was customary to deny that Rockefeller had any role in the operation of Standard Oil. But confidential correspondence in the Rockefeller archive shows that this was indeed not the case and clearly supports the views expressed by Gibb. As Rockefeller himself wrote:

> Although I have been for a quarter of a century out of active connection with the business, I have kept informed and been generally familiar with its current progress ... I need not give here expression to my confidence in the administration since the discontinuance of my active relations with the company.[85]

Moreover, when John D. Rockefeller's grandson, Nelson, passed through the procedures for his confirmation as vice-president, he was obliged to publish data on Standard Oil stocks that he owned and noted that the Rockefeller family "has historically been associated" with the company.[86]

The obstruction, which the oil magnates arranged at the Genoa Conference, had a harmful effect both on the fate of the negotiations more generally and on the possibility of a Soviet-American dialogue. Immediately after Genoa, the Standard Oil representative (Mowinkel) expressed his satisfaction that the conference ended without any "disagreeable surprises."[87]

But Genoa was also a defeat for those who made it their goal to impose their own conditions on Soviet Russia. When the Soviet delegation set off for the Genoa Conference, Lenin observed that the Soviet representatives are going there like merchants, that they are prepared to do business with the Western countries and to make agreements on mutually beneficial forms of economic relations. Lenin warned that the Soviet Union will not negotiate on any other terms. "If Messrs. capitalists think that they can procrastinate and the longer, the greater the concessions," said Lenin, "then I repeat, it is necessary to tell them: "*That's enough—tomorrow you shall get nothing*!"[88]

The Genoa Conference, like the later conference in The Hague (June 15, to July 19, 1922), ended in failure. One of the most important lessons of these events was the fact that the Western powers could not dictate terms and conditions to Soviet Russia.

After Genoa and The Hague, Deterding assured the Americans that he was not thinking of a separate peace with the Bolsheviks. However, the British secret service learned from the Dutch police that an agent of Royal Dutch Shell had attempted to make contact with the Soviet representatives in order to circumvent Standard Oil and obtain oil concessions in Russia.[89] News of this reached the Americans as well. Deterding hastened to deny the reports. On July 18, 1922 he met Teagle in London and proposed that they sign an agreement for joint operations.[90] After consultations with the State Department, Teagle signed an agreement with Deterding and Nobel on July 24, 1922 to create a "united front" against Soviet Russia. At a meeting specially convoked in Paris in September 1922, the leaders of the largest oil companies considered concrete measures to implement the above agreement. Rockefeller's representatives, however, endeavored to avoid assuming excessive obligations, fearing that Deterding would somehow exploit these in the future. "We ought to avoid at present," wrote one of the Standard Oil leaders, "any open initiative from which blame could be fixed on us to the effect that we are at the bottom of the opposition against Russia."[91] This was all the more important since another American company, Sinclair, had already concluded an advantageous agreement with the Soviet government.

Sinclair's actions affected the interests of Standard Oil and forced the company's leadership to reconsider its position. The behavior of Western European diplomats also aroused concern, as they began to extend diplomatic recognition to Soviet Russia; England, followed by Italy, France, and other countries, reestablished diplomatic relations with the USSR. And, in violation of his obligations for a "united front," Deterding made a deal with Soviet trade organizations to purchase large quantities of oil.

The Americans had plainly lost. In early 1923 Teagle, with regard to the purchase of the Nobel enterprises, complained that it was very much like "sitting up with a sick child" and "nursing it along for years." "Please don't think I am a pessimist, for such is not the case," he wrote Heinrich Riedemann, a Standard Oil leader in Germany, "but I am not entirely satisfied in my mind that some of our recent investments, particularly in foreign producing possibilities, can be in every way justified." Riedemann shared that view and assumed that Standard Oil should quickly change its tactics and repudiate its earlier aims. "Certainly the situation in Russia is very abnormal and represents an unprecedented case. The participation of a government in industrial and business enterprise as in Russia is new and unheard of in the history of business." But he counselled flexibility: "None of us like the thought of helping Soviet ideas, but if others should be willing to come in, what would then have been the use if we had kept aloof?"

With each passing day this question became more and more disturbing to the leaders of the American company. They consistently avoided contacts with the Soviet side, but their partners in the "united front" had obviously deserted them. "I am really afraid that we shall have to swallow the distasteful pill and accept the dictates of the present situation." True, Teagle still hesitated, wondering whether Standard will not play into the "Reds'" hands and whether this will not be an encouragement to "irresponsible governments" elsewhere. Nevertheless, the circumstances forced the company to take urgent measures. "Man is a strange being," said Riedemann, "and in spite of all disappointments, he still starts every year with new hopes. So let us do the same." His proposal was adopted.

Beginning in 1924 Standard Oil began to make massive purchases of Soviet oil. American representatives endeavored even to reach an agreement for several years' advance purchase of the entire oil production of the USSR. This they were refused. But the regular sales increased in volume year after year. The American activity increased especially after the relations between the USSR and England were severed in 1927. Significantly, it was precisely at this point that one of the most influential advisors in the Rockefeller company, Ivy Lee, proposed that the United States recognize the USSR and even made a trip to the Soviet Union.

Upon Lee's return to America, F. Powell (the head of the Anglo-American Company, a British affiliate of Standard Oil) said: "I am extremely anxious to hear your account of what you have seen and done while in Russia." "What I should like to know," he continued, "is whether the stability of the present government is increasing or decreasing."[93] The fact is that hopes for the collapse of the Soviet state were not justified, and there were no signs of its growing weaker.

In a memorandum to the State Department, Ivy Lee proposed to begin conversations for the establishment of diplomatic relations and simultaneously to resolve misunderstandings in trade between two such great countries as Russia and the United States.[94] In his book, *Present-Day Russia,* published in 1928, Lee proposed to move from verbal propaganda to concrete action. "We have tried to isolate, and other countries sought to resort to armed intervention. But this did not cure the disease of Bolshevism. What then remains?"[95] Lee strongly recommended that the United States reconsider its relations with the Soviet Union. "The main thing," he wrote in a message to the New York Chamber of Commerce, "is that American business should establish contact with the Russian government and strive to ensure the development of business relations between the two countries."[96]

But even this quite moderate appeal, with many reservations, provoked furious attacks. "What are you doing all this for?" asked the *Wall Street Journal,* "Who is paying you for it?" The absolutely absurd rumor was spread that the Soviet government had allegedly hired Lee to conduct propaganda in favor of diplomatic recognition. A document alleging that Lee received money from the Soviet Commissariat of Foreign Affairs was fabricated and given to American intelligence services; the falsification was easily refuted.[97]

Significantly, one newspaper, *The Washington Star,* observed that "neither Standard Oil nor any other company, whose interests Mr. Lee represents, has ever addressed the State Department on the question of recognizing the USSR."[98] Moreover, Teagle continued to oppose proposals to normalize relations with the Soviet Union. Calling Ivy Lee's appeals "poison," he himself in so many respects was backing away from the irreconcilable goals he had held earlier.

In discussing his meeting with Teagle, Gibb recounts that the American magnate loved Deterding. "They thought in the identical categories and had identical tastes. They were very close to the management of the petroleum world and would have doubtless been even closer if it had not been for the refractoriness of the Russians, who refused to "play the game."[99] Indeed, much united Teagle and Deterding, but a constant tension still divided the two. Gulbenkian, who knew both men well, called Teagle an "upstart" and Deterding "a turncoat." Reflecting the growing appetites of American business, the "upstart" Teagle impetuously strained forward. Standard Oil seized new markets and oil deposits, and he was the soul and direct executor of this expansionist policy.

As for Deterding, he was also an experienced oilman with just one difference: his actions bore a more refined character. He thought nothing about deserting and changing sides. Deterding "the turncoat" was a difficult partner. He constantly infuriated Teagle by violating agreements.

Like Teagle, Deterding wished to see the Soviet system collapse. His relationship to Soviet Russia amounted to little more than profound hatred. In September 1920, at the height of the armed intervention, Deterding expressed the hope that "not only the Caucasus, but all Russia will be purged of the Bolsheviks within in about a half year." Even after the failure of intervention, Deterding was still asserting at the end of 1925 that "in less than a year Russia will again become a civilized country"—which, in his mind, meant a return to capitalism.[100] He preserved his hostility to the Soviet Union to the end of his life. But when it was profitable, Deterding made deals with the USSR, thereby making fools of his allies and abandoning their agreed-upon policy. The actions of Royal Dutch Shell forced the Americans to rethink their relations to the Soviet Union. "We have doubts as to what course it would be better to follow," declared one leader of Standard Oil shortly before Lee's trip to the Soviet Union.

After his visit to the USSR in 1927, Lee compiled a report on the results of his trip. The working catalogue of the Rockefeller headquarters in New York, which includes the rubric "Russia-USSR," has a reference to the Ivy Lee report, but the document itself is not in the archive; it has been lost, I was told. In any event, upon Lee's initiative and taking into account Rockefeller's interest in the oil business and realistically assessing the world situation, the firm began to reconsider the question of normalizing the entire complex of relations between the United States and the Soviet Union. Six years would pass before the United States officially recognized the USSR, but movement in that direction had already become irreversible.

An interview published by *Le Figaro* in March 1986 with David Rockefeller (the present head of the Rockefeller dynasty and former president of the Chase Manhattan Bank) made clear that Ivy Lee's activities were sanctioned by the Rockefellers. Beginning in 1927, the oil corporations and the Chase National Bank—all under their control—began to make regular business contacts with the Soviet Union. A step was thus taken in the direction of changing the relationship to the USSR. "The influence of the group, Chase National Bank, which is conducting trade with the USSR," wrote the American journalist Scott Nearing in 1927, "undoubtedly will give the upperhand [over opponents of establishing relations], and after this will come recognition of the USSR."[101] Progress in the question of normalizing relations with the USSR, evidently, was assisted by the fact that the Soviet Union had reestablished its position on the world petroleum market at a rapid tempo.

In oil production and exports, Soviet Russia was behind America. But the Soviets delivered to the world market millions of tons and acted independently of the big corporations, selling oil at lower prices. Discussing possible relations with the Soviets, the Standard Oil Company's representatives hoped to establish some kind of stabilization. In reply to Ivy Lee's letter from Moscow, F. Powell wrote that he followed "with a great deal of interest your movements in Russia" and was looking forward to seeing Lee on his way back in London. "Here the oil situation is still acute," wrote Powell, "owing to the continuance of the very annoying Russian competition at low prices. Our competition absolutely declines to move on markets here unless the Russian competition can be controlled in some way or unless they advance their prices simultaneously with our own."[102] Several years passed before a more or less satisfactory settlement was reached. This occurred in the early 1930s, when the situation in the world market radically changed with the decrease in Soviet oil exports and a general fall in prices, which by then coincided with American diplomatic recognition of the Soviet Union.

Meanwhile, in the early 1920s, the United States first experienced a petroleum famine: severe gasoline shortages hit the state of California. In the opinion of two American scholars (A. Olmstead and P. Rhode), this was an early warning of the future world energy crisis of the 1970s.[103] In any event, in July 1920, the flow of automobiles on the roadways of California suddenly died, as gasoline pumps were closed and J. D. Rockefeller Jr. was forced to interrupt his holidays and return to New York. Although the gasoline famine of 1920 was soon overcome, the episode provided a strong impulse not only for finding foreign sources of oil and but also for making purchases of Soviet oil.

Immediately after the expulsion of foreign troops from the territory of the Soviet Union, efforts were begun to reestablish the oil industry. The first steps were taken in 1921, and by the mid-1920s tangible results had already been achieved. In 1924-1925 investments in the Soviet petroleum industry amounted to 107 million gold rubles, which represented 27.7 percent of all industrial capital investment in the USSR. Oil exports brought in a noticeable income. This money was used to acquire foreign equipment, which was then used to reconstruct the oil industry, thereby promoting its rapid growth. In 1927 the level of oil output exceeded prewar production by 10 percent, and the tempo of development of oil exports proved still more rapid. Whereas in 1913 Russia exported 951,000 tons, by 1926-27 it shipped abroad 1.9 million tons and in 1932 5.6 million tons—that is, 6.5 times more than before World War I.[104] In addition, it should be emphasized that Soviet Russia became one of the greatest exporters of oil products. In 1933 it held fourth place in the world for gasoline exports and third in the sale of kerosene and lubricants—despite the boycott of Western firms, which had lasted a long time. The Soviet oil syndicate Soiuznefteeksport firmly established itself on the world market, competed successfully with the powerful corporations and concluded profitable deals with them. A reduction of Soviet oil exports commenced in 1933, both because of the falling world prices for oil and because of the growing demand for oil products in the USSR—as a result of industrialization and the increased production of motor vehicles, tractors, and combines, all of which demanded ever larger quantities of fuel.[105]

Thus, despite the serious destruction during the war and foreign intervention, the petroleum industry in Russia was reestablished in slightly more than a decade and experienced an major boom. The Soviet oil export proved an important factor on the world market. History has shown that this factor was fated to acquire still greater significance in the future.

NOTES AND REFERENCES

1. K. V. Gagelin [Hagelin], *Moi trudovoi put'* (New York, 1945), pp. 342-343 (Hagelin was the first to relate his conversation with Gwinner); Tolf, p. 194.
2. H. Williamson, R. Andreano, A. Daum, and G. Close, *The American Petroleum Industry. The Age of Energy 1899-1959* (Evanston, 1963), *2,* 268.
3. Jones, p. 178.
4. L. H. Seltzer, *A Financial History of the American Automobile Industry* (Boston, New York, 1928), pp. 75-81; J. S. Clark,. *The Oil Century* (Norman, 1958), p. 125.
5. C. Tugendhat, *The Biggest Business* (New York, 1968), p. 71.
6. U. Brack, *Deutsche Erdölpolitik vor 1914* (Hamburg, 1977), pp. 504-506.
7. Telegram to Sazonov (June 16/29, 1914). In *AVPR,* f. Kantseliariia, 1914 g., d. 280, l. 2.
8. F. Fischer, *Der Griff nach der Weltmacht* (Düsseldorf, 1962), p. 112.
9. Pravlenie Astra-Romana—poslannikam Rossii, Anglii i Frantsii (November 15/28, 1914): Archiva "Muntenia," Astra-Romana. Konfidentsial'-naia perepiska, d. 392/2, l. 210; G. Ravash, *Iz istorii rumynskoi nefti* (Mowcow, 1958), p. 75.
10. Ravash, pp. 75, 78-79.
11. Ibid., p. 81.
12. G. B. Clemenceau to W. Wilson (December 15, 1917). In H. Berenger, *Le petrole et la France* (Paris, 1920), pp. 59-60.
13. Williamson, et al., vol. 1, ch. 8.
14. *Ekonomicheskoe polozhenie Rossii nakanune Velikoi Oktiabr'skoi sotsialisticheskoi Revoliutsii. Dokumenty i materialy* (Moscow-Leningrad, 1957), *2,* pp. 627-628; *Osobye soveshchaniia i komitety voennogo vremeni* (Petrograd, 1917), pp. 19-23.
15. Reports of agent "Y" (November 9, and March 17, 1909). In United States National Archives, U.S. Office of Naval Intelligence, RG 38, Box 838, File 2080.
16. Memorandum from I. K. Grigorovich (September 4/17, 1912). In *TsGAVMF SSSR,* f. 418, op. 1, d. 2645, l. 15.
17. G. H. Melville (Rear Admiral, Engineer-in-Chief, Bureau of Steam Engineering, U.S. Navy Department) to D. T. Mertvago (September 13 and 15, 1899) in the Hoover Institution Archives, f. Posol'stvo Rossii v Vashingtone, Perepiska voenno-morskogo agenta.
18. Lenin, *37,* p. 50.
19. See: F. Landberg, *60 semeistv Ameriki* (Moscow, 1948), pp. 525-526; G. S. Gibb and E. H. Knowlton, *The Resurgent Years, 1911-1927* (New York, 1956), p. 598.

20. Gibb and Knowlton, pp. 107-8.
21. V. Vol'skii, *Latinskaia Amerika, neft' i nezavisimost'* (Moscow, 1964), p. 50.
22. Dispatch from A. P. Kassini (January 17/30, 1901). In *AVPR,* f. Politarkhiv, 1901 g., d. 1001, l. 90.
23. Gibb and Knowlton, p. 182, 195-197.
24. Ibid., pp. 197-199.
25. P. Foley, "Petroleum Problems of the World War," *U.S. Naval Institute Proceedings,* November 1924, p. 1821.
26. Jones, pp. 182, 185.
27. Ibid., p. 189.
28. P. Cambon to G. Dumergue (April 27, 1914). In *Documents diplomatiques français,* Ser. 3, t. 10 (Paris, 1936): 430.
29. M. Kent, *Oil and Empire. British Policy and Mesopotamian Oil, 1900-1920* (London, 1976), pp. 120-124.
30. Gibb and Knowlton, p. 108.
31. A. Bedford to R. Lansing (December 21, 1921), and Bedford's memorandum to the State Department (December 1, 1921) in the Library of Congress, Lansing Papers.
32. Instruction of the State Department (June 16, 1919). In U.S. National Archives, State Department Files.
33. Telegram sent by Lansing on December 8, 1919. In U.S. National Archives, State Department Files.
34. Memorandum of Bedford to the State Department (December 1, 1921). In the Lansing papers.
35. Memorandum of October 30, 1919 in the U.S. National Archives, State Department files.
36. J. Davis to Hughes (May 18, 19200. In ibid.
37. Tugendhat, p. 75.
38. Seltzer, p. 76.
39. Kent, pp. 153-154.
40. Board of Standard Oil to the State Department (May 28, 1920). In the U.S. National Archives, State Department files.
41. Minutes of the meeting held August 3, 1920 in ibid.
42. Colby to Curzon (November 20, 1920) in ibid.
43. Statement by F. E. Powell (chairman of the Anglo-American Company, an affiliate of the Standard Oil Company), as quoted in the *Sunday Times,* December 12, 1920.
44. A. Ledu to H. Hoover (April 13, 1921). In U.S. National Archives, State Department files.
45. See Van H. Manning to C. E. Hughes (August 10, 1921). in ibid.
46. A. Dulles to Lansing, February 15, 1922. In the Library of Congress, Lansing papers.
47. Dulles to Lansing (February 15, 1922). Lansing papers.
48. W. Fert to Manning (November 21, 1921). In U.S. National Archives, State Department files.
49. Tugendhat, p. 82.

50. N. Rhodes to J. Garfield (27.V.1914) in the Library of Congress, Garfield Papers.

51. Garfield to Lansing (July 5, 1915). In ibid.

52. Rhodes to Garfield (July 1, 1916). In ibid.

53. *Daily Telegraph,* December 11, 1920.

54. Gibb and Knowlton, pp. 363-364.

55. Ibid., pp. 394-391.

56. Memorandum of the Foreign Trade Division to the Assistant Secretary of State, F. M. Dearing (May 10, 1921) in the U.S. National Archives, State Department files.

57. Ibid.

58. Gibb and Knowlton, pp. 388-389.

59. Memorandum of the State Department section on Latin America (February 16, 1924) in the U.S. National Archives, State Department files.

60. M. Nicolson to K. Hale (June 29, 1935). In ibid.

61. Memorandum from Lansing (February 16, 1922) in the Lansing papers.

62. *El Universal,* November 30, 1922 (clipping in U.S. National Archives, State Department files; N. A. Graebner (ed.) *Uncertain Tradition; American Secretaries of State in the Twentieth Century* (New York, 1961), p. 133.

63. Memorandum of M. Requa and N. Bicher (May 12, 1919). In the U.S. National Archives, State Department files.

64. Memorandum of the juridical section of the State Department to Hughes (May 1, 1923). In the U.S. National Archices, State Department files; memorandum of the Solicitor to the Secretary of State (May 16, 1921), Note of April 20, 1923, in ibid.

65. Gibb and Knowlton, p. 347.

66. Ibid., p. 347.

67. Ibid., pp. 331-333.

68. Bedford to Hughes (March 10, 1922). In U.S. National Archives, State Department files.

69. Gibb and Knowlton, p. 333.

70. W. Teagle to Lansing (March 19, 1920). In U.S. National Archives, State Department files.

71. Hughes to Child (May 2, 1922). In U.S. National Archives, State Department files.

72. A. C. Fink, *The Genoa Conference: European Diplomacy, 1921-1922* (London, 1984), pp. 225-229.

73. Memorandum (dated December 1920), in U.S. National Archives, State Department files.

74. Minutes of the meeting January 12, 1921. In U.S. National Archives, State Department files.

75. V. A. Shishkin, *Sovetskoe gosudarstvo i strany Zapada v 1917-1923 gg.* (Leningrad, 1969), pp. 177-191, 195, 249-257, 285-287.

76. Bedford to Hughes (March 10, and March 5, 1922). In U.S. National Archives, State Department files.

77. Bedford to Hughes (May 11, 1922). In U.S. National Archives, State Department files.

78. Krassin to Chicherin (May 29, and June 11, 1920). In *Dokumenty vneshnei politiki SSR* (Moscow, 1958), 2, p. 552, 569.

79. Memorandum on the contents of a telephone conversation with Bedford on January 31, 1922, in the U.S. National Archives, State Department files.

80. Bedford to Hughes (May 11, 1922). In U.S. National Archives, State Department files.

81. G. Gibb to A. A. Fursenko, September 14, 1969 and November 5, 1969.

82. S. V. Zarnitskii and L. I. Trofimova, *Sovetskoi strany diplomat* (Moscow, 1958), p. 150.

83. Gibb to Fursenko, September 14, 1969.

84. Gibb to Fursenko, December 17 and 30, 1968.

85. J. D. Rockefeller to A. Bedford (January 1, 1920) *RAC*.

86. Statement of Nelson A. Rockefeller to the Committee on Rules and Administration of the United States Senate, September 23, 1974, pp. 17-18 (news release). Copy in *RAC*.

87. Gibb and Knowlton, p. 339.

88. Lenin, *45*, p. 2, 13.

89. Report of the Intelligence Service (July 20, 1922). In the *Public Record Office*, Foreign Office, 371/8162.

90. Teagle to Bedford (July 18, 1922). In the U.S. National Archives, State Department files.

91. W. Riedemann to A. Bedford (September 17, 1922). In Gibb and Knowlton, p. 341.

92. Ibid., p. 343, 346, 348, 351.

93. F. Powell to Ivy Lee (August 21, 1928). In the Ivy Lee Papers (Princeton University; Princeton, NJ).

94. Ivy Lee to F. Kellog (July 8, 1927) in ibid.

95. Ivy Lee, *Present-day Russia* (New York, 1928), pp. 153-156.

96. Ivy Lee to Pearson (July 5, 1927) in the Ivy Lee Papers, Princeton University.

97. R. E. Hiebert, *Courtier to the Crowd* (Ames, 1966), p. 280f.

98. *Washington Star*, June 1, 1926.

99. Gibb to Fursenko, December 30, 1968.

100. G. Jones and C. Trebilcock, "Russian Industry and British Business, 1910-1930: Oil and Armaments," *Journal of European Economic History*, (Spring 1982) *11*, pp. 89-90.

101. S. Niring [Nearing], "Vneshniaia politika Soedinennykh Shtatov. Doklad v MGU 23.XI.1927," *Mirovoe khoziaistvo i mirovaia politika*, 1928, *1*, p. 59.

102. Power to Ivy Lee (August 21, 1928) in the Ivy Lee Papers (Princeton University).

103. A. L. Olmstead and P. Rhode, "The U.S. Energy Crisis of 1920 and the Search for New Oil Supplies," in *Oil in the World Economy; the Ninth International Economic History Congress*, edited by R. W. Ferrier and A. Fursenko (Bern, 1986), pp. 23-33.

104. See Tables 5 and 6 in the Appendix.

105. V. A. Shishkin, "Soviet Oil Exports between the Two World Wars," in *Oil in the World Economy*, pp. 14-22.

Conclusion

Toward the end of the 1920s a new era of the battle for oil had begun, bringing further strains and tensions in the international competition for petroleum. This was due to a variety of factors, above all the increasing demand for oil, which had acquired exceptional importance as an energy source and was used more and more as a raw material in the chemical industry. The conversion of naval vessels and commercial shipping to engines using oil, the dissemination of oil-consuming equipment in industry, the explosive development of land transport (especially motor vehicles) using gasoline and other forms of petroleum products)—all that transformed oil into a vitally important source of energy. The colossal growth in the significance of oil fuel and the sharp rise in demand for it caused the struggle for sources of oil to attain an unprecedented scale. The development of entire branches of industry and transport proved dependent upon obtaining oil. Their progress, and the very viability of many countries, depended upon oil deliveries. All this meant that oil corporations, which had earlier relied upon government support, now openly collaborated with home governments in their campaigns to acquire the most promising oil deposits around the globe.

During three decades, from 1900 to 1930, the world production of oil increased ten-fold. Before 1900, the main producers and exporters of oil products on the world market were Russia and the United States. The other oil-producing countries—Rumania, Austria-Hungary, Dutch Indies, Burma, and Canada—accounted for only an insignificant share of world output. Then the situation began to change. In 1913 the United States produced 33 million tons, Russia 8.7 million, Mexico 3.4 million, Rumania 1.8 million, and the Dutch Indies 1.5 million.[1] The ensuing rapid growth of oil output during World War I and in the postwar period occurred not only as a result of more intensive exploitation of old petroleum deposits, but also through the exploitation of new reserves, above all in the Near East (Iran and Iraq), as well as in Latin America (Mexico and Venezuela).

At the very end of the nineteenth century, in terms of the quantity of oil products used and their proportion of all forms of energy consumption, America and Europe occupied approximately an equal position—approximately 4 percent. However, by 1930 the United States already used oil to cover 25 percent of its energy needs, while Europe would reach this level only some 20 years later.[2]

The British journalist H. O'Connor, author of the book *Empire of Oil,* observed that even in the mid-twentieth century an enormous number of dwellings were still using kerosene for lighting. In the early twentieth century, kerosene comprised approximately two-thirds of the oil production in the United States. In America itself, the kerosene lamp and heating devices were gradually replaced by gas or electrical lamps and heating systems. Nevertheless, during the first two decades of this century, the quantity of kerosene used in the United States rose 2.5 times, and the number of kerosene lamps still amounted to 20,000,000 units.[3] By the early twentieth century, only individual vessels had been converted to use oil fuel. Even on the eve of World War I, just 3 percent of the ships were using oil. But by 1929, through the entire world, 30 percent of the ocean-going ships were using oil for fuel. By the end of the period examined here, the world's vehicular pool included tens of millions of busses, automobiles, and trucks, which each year consumed millions of tons of gasoline.

Thanks to the intensive exploitation of oil, above all in the newly discovered deposits in the Near East and Latin America, the world market was saturated with sufficient quantities of oil products to satisfy the growing demand. An important factor in ensuring a sufficient supply of oil for the world economy was the export of Soviet oil. As indicated, it was outside the sphere of influence of corporations seeking to monopolize the oil trade and to rack up higher prices. In this sense, Soviet oil exports from the very beginning played a stabilizing role on the world market because of the independent, low prices followed in Soviet export policy.

The further evolution of the battle for oil proved to be intimately tied to the development of new forms of finance capital, with a much closer intertwining of interests and interaction of industrial monopolies and banks. In the period between the first and second world wars, the struggle between the largest corporations to gain control over oil deposits and markets continued to intensify. World War II furthered this intesity, and afterwards virtually the entire world market was divided among the "Big Seven"—the seven largest multinational corporations, which had established control over an enormous part of the production and sale of oil productions in capitalist countries.

This concerns the five corporations of American origin: Exxon (Standard Oil of New Jersey), Mobil (Standard Oil of New York), Standard Oil of California (Chevron), Texaco (Texas Oil Co.) and Gulf (Gulf Refining Co.), to which are added two corporations of British origin: British Petroleum Co. and the Anglo-Dutch Royal Dutch Shell. The general sum of the assets of these firms in the early 1970s, according to minimal estimates, amounted to 132 billion dollars. As already noted, in 1984 one of the "seven sisters" of the International Oil Cartel swallowed another "sister" (Gulf), thereby reducing the "Seven" to "Six." The oil business also represented the most advanced form of capitalist entrepreneurship in the degree of concentration of production and capital. And it has preserved this characteristic to the present day. Truly, oil is "the biggest business."[4]

The significance and scale of the oil business is defined, to a considerable degree, by the fact that not only has it preserved its international political character, as in the period examined here, but also acquired still greater significance for the future. For example, in the period of the energy crisis of the 1970s, the problem of supplying oil, attempts of trans-national corporations (and the countries in which they base themselves) to maintain control over strategically important sources of oil led to an unprecedented level of international tension, wars and conflicts, which were fraught with the danger of a new world war. Earlier, politics was already closely linked with the actions of the oil corporations. But in our time this phenomenon has become still more marked, although the preconditions for this were laid precisely in the course of the battle for oil at an earlier stage of international oil competition.

NOTES AND REFERENCES

1. See Table 5 in the Appendix.
2. Jones, p. 5.
3. Ibid., p. 4.
4. R. Engler, *The Brotherhood of Oil. Energy Policy and the Public Interest* (Chicago, 1977), p. 19.

Abbreviations

Archiva "Muntenia" A.R.	Archiva societatii petrolifere "Muntenia" Astra-Romana
AVPR	Arkhiv vneshnei politiki Rossii Ministerstva inostrannykh del SSSR
Documents Diplomatiques	*Documents diplomatiques français relatifs aux origines de la Guerre de 1914*. Paris.
LGIA	Leningradskii gosudarstvennyi istoricheskii arkhiv
IRLI	Institute russkoi literatury AN SSSR (Pushkinskii dom)
LOII	Leningradskoe otdelenie Instituta istorii AN SSSR
MKNPR	*Monopolisticheskii kapital v neftianoi promyshlennosti Rossii. 1883-1894. Dokumenty i materialy*. Moscow, Leningrad, 1961.
Moniteur	*Moniteur du petrole Roumain*. Bucharest.

Parliamentary Debates	*Hansard's Parliamentary Debates.* House of Commons. London
RAC	Rockefeller Archive Center
Reichstag	*Stenographsche Berichte über die Verhandlungen des Deutschen Reichstages.* Berlin.
TsGAVMF SSSR	Tsentral'nyi gosudarstvennyi arkhiv Voenno-Morskogo flota SSSR
TsGIA AzSSR	Tsentral'nyi gosudarstvennyi istoricheskii arkhiv Azerbaidzhanskoi SSR
TsGIA GSSR	Tsentral'nyi gosudarstvennyi istoricheskii arkhiv Gruzinskoi SSR
TsGIA SSSR	Tsentral'nyi gosudarstvennyi istoricheskii arkhiv SSSR
US Commercial Relations	*Commercial Relations of the United States.* Washington
US Consular Reports	*Consular Reports of the United States.* Washington
US Mineral	Mineral Resources of the United States

Appendix

Table 1. Volume of Trade of the Nobel Brothers Co. 1893-1913 (Thousands of Rubles)

Year	*Fuel Oil*	*Kerosene*
1893	638.5	292.3
1894	782.2	377.7
1895	844.8	421.9
1896	1068.1	412.7
1897	1099.2	433.5
1898	1171.2	470.9
1899	1350.1	647.0
1900	1560.0	587.2
1901	1474.6	572.3
1902	1441.7	665.2
1903	1674.4	588.4
1904	1820.4	588.3
1905	1702.7	557.2
1906	1441.1	484.1
1907	1365.3	515.8
1908	1491.7	685.8
1909	1405.3	625.6
1910	1492.3	685.8
1911	1726.9	700.6
1912	1415.3	655.7
1913	1314.0	623.5

Source: *MKNPR,* pp. 753-754.

Table 2. Income from the Nobel Brothers' Corporation. 1883-1913 (thousands of rubles)

Year	*Fuel Oil*	*Kerosene*
1883	1740	5130
1893	5540	14641
1895	7794	19859
1899	17706	31913
1900	26439	36205
1905	28223	33761
1908	26856	40821
1909	29282	42675
1911	30034	43812
1913	39420	52479

Source: MKNPR, pp. 753-754.

Table 3. Profits and Divides of the Nobel Brothers' Corp. 1879-1914 (thousands of rubles)

Year	*Net Profit*	*Dividends*
1879	128	—
1880	1005	20
1881	1443	17
1882	1770	18
1883	2072	15
1884	1793	10
1885	859	—
1886	—	—
1887	1755	6
1888	1658	6
1889	2499	6
1890	2337	8
1891	1451	5
1892	1484	6
1893	612	5
1894	1795	6
1895	2171	10
1896	2032	10
1897	1432	10
1898	2020	10
1899	4482	18
1900	6911	20
1901	3833	20
1902	2159	15
1903	2730	10

(continued)

Table 3. (continued)

Year	*Net Profit*	*Dividends*
1904	2232	12
1905	2971	10
1906	5480	12
1907	5653	20
1908	4151	20
1909	2978	12
1910	352	12
1911	3499	14
1912	11238	22
1913	14695	26
1914	13142	26

Source: *MKNPR,* p. 750.

Table 4. Kerosene Prices of the Nobel Brothers' Corp. (kopeck price per pood)

Year	*Cost Price of Manufacturing at Nobel Plant*	*Purchase Price from other firms*	*Sale Price in Russia*	*Sale Price Abroad*
1890	10.0	28	88	55
1891	9.8	44	85	45
1892	7.2	37	87	40
1893	7.4	63	90	20
1894	6.9	36	89	11
1895	9.0	70	90	21
1896	10.9	70	94	22
1897	—	90	91	21
1898	8.3	70	90	25
1899	6.4	64	100	38
1900	11.3	80	114	51
1901	8.2	65	100	35
1902	7.2	54	100	30
1903	12.1	63	105	—
1904	15.2	70	112	48
1905	19.2	95	102	42
1906	24.1	100	117	53
1907	27.4	70	135	60
1908	24.8	60	122	52
1909	28.8	60	121	50
1911	22.7	52	114	45
1912	29.1	65	133	59

Source: *MKNPR,* p. 753.

Table 5. Oil Production
(thousands of tons)

Year	*Russia*	*USA*	*Romania*	*Austria-Hungary*	*Germany*	*Indonesia*	*Others*	*Total*
1881	480	3684	16	38	3		39	4260
1882	604	4042	18	44	8		39	4755
1883	799	3123	19	49	3		36	4029
1884	1439	3225	28	54	6		38	4790
1885	1854	2911	26	62	5		38	4896
1886	2398	3737	22	41	10		83	6291
1887	2446	3766	24	46	10		74	6366
1888	3069	3677	29	62	11		99	6947
1889	3277	4683	40	69	9		113	8191
1890	3821	6102	51	88	14		129	10205
1891	4604	7230	64	84	14		136	12132
1892	4764	6727	79	86	13		1480	13149
1893	5388	6449	71	92	13	80	164	12257
1894	4844	6571	68	126	16	92	180	11897
1895	6144	7044	77	193	16	162	169	13805
1896	6288	8118	72	325	19	190	180	15202
1897	7244	8053	76	296	22	340	209	16240
1898	8204	7373	103	316	24	395	221	16636
1899	8783	7600	190	308	26	239	318	17464
1900	10091	8472	217	312	48	300	420	19860

Year	*Russia/ USSR*	*USA*	*Rumania*	*Austria-Hungary/ Poland*	*Germany*	*Indonesia*	*Mexico*	*Others*	*Total*
1901	11342	9240	223	433	42	534	1	482	22297
1902	10725	11821	274	552	47	324	5	470	24238
1903	10066	13378	368	697	59	768	10	606	25952
1904	10459	15591	479	792	65	867	17	734	29024
1905	7319	17940	589	768	79	1045	33	466	28644
1906	7843	16845	849	728	77	1089	67	902	28400
1907	8237	22119	1081	1126	101	1329	134	1024	35151
1908	8281	23774	1099	1680	134	1369	524	1130	37991
1909	8785	24393	1242	1988	136	1470	361	1404	39779
1910	9367	27907	1295	1688	137	1469	484	901	43648
1911	8814	29357	1479	1401	135	1621	1672	1379	45838
1912	9058	29688	1728	1137	137	1444	2205	1538	46935
1913	8767	33085	1801	1041	114	1488	3422	1598	51316
1914	8925	35391	1708	670	104	1535	3494	2445	54272
1915	9128	37434	1602	554	94	1598	4383	2741	57534
1916	9695	40053	1191	860	87	1671	5302	2066	60295
1917	9316	44654	495	794	85	1651	7363	2612	66970
1918	5388	47399	1163	745	36	1703	8500	2119	67053
1919	4566	50387	868	833	35	2023	11600	3714	74026
1920	3386	58984	990	747	33	2344	20917	4337	91738
1921	3991	62880	1114	688	36	2258	25755	5277	101999
1922	4753	74246	1311	696	42	2273	24274	6784	114379
1923	5213	97534	1447	719	46	2646	19920	7740	135265
1924	6040	95075	1780	753	54	2726	18601	10047	135076
1925	6984	101707	2217	794	72	2852	15383	12342	142351
1926	8564	102657	3105	778	87	2829	12041	16011	146072
1927	10256	120003	3511	711	88	3548	8539	21370	167936

Source: This table was compiled by S.A. Isaev from data in the annual publications of the United States Mineral Resources and from materials in official Soviet statistics.

Table 6. Oil Exports
(thousands of tons)

	Russia			United States		
Year	*Oil*	*Kerosene*	*Basic Oil Products*	*Oil*	*Kerosene*	*Basic Oil Products*
1881	—	10	16	128	1403	1484
1882	—	11	18	142	1352	1433
1883	—	40	57	186	1389	1476
1884	—	76	113	251	1369	1482
1885	18	119	177	257	1407	1495
1886	18	20	246	241	1531	1621
1887	18	20	311	255	1531	1635
1888	1	457	580	245	1381	1556
1889	—	560	719	269	1530	1873
1890	12	676	800	305	1594	1879
1891	194	737	887	305	1740	1818
1892	195	801	937	329	1722	2019
1893	210	819	971	346	2126	2280
1894	280	709	867	344	2190	2352
1895	246	839	1040	349	2043	2221
1896	415	860	1043	355	2259	2452
1897	386	386	857	367	2400	2599
1898	715	715	921	346	2295	2529
1899	398	398	1130	353	2156	2432
1900	635	635	1233	414	2197	2480
1901	1353	1289	1505	410	2490	3192
1902	684	1387	1635	475	2342	3109

1903	531	1813	1966	491	2080	2899
1904	909	1841	2146	360	2293	3341
1905	686	727	942	410	2653	3489
1906	670	618	785	390	2735	3394
1907	718	554	734	399	2817	3109
1908	903	609	824	471	3276	4769
1909	1025	575	801	538	3301	4928
1910	1038	514	857	568	2966	4794
1911	959	1386	1680	637	3509	3593
1912	1013	1417	1721	596	3238	5993
1913	739	440	951	614	3531	6792
1914	1151	1382	1658	394	3187	7191
1915	1452	no data	no data	499	2641	7343
1916	no data	no data	no data	543	2698	8257
1917	no data	no data	no data	543	no data	8390
1918	39	0	2	650	1574	8656
1919	0	0	0	802	3133	7857
1920	0	0	0	1116	2780	9988
1921	0	0	7	1191	2399	8922
1922	1	2	51	1353	2862	17046
1923	85	3	49	2315	2810	22639
1924	74	335	637	2394	2925	12917
1925	63	454	1309	1776	2825	13077
1926	117	344	1357		2873	15461
1927	151	474	1925		2780	16659

Source: This table was compiled by S.A. Isaev on the basis of the annual publications of the United States Mineral Resources as well as the materials in official Soviet statistics.

Bibliography

Archival Sources

Archiva Societatii Petrolifere "Muntenia" (Bucharest)
Collection: Astra Romana
AVNP: Arkhiv Vneshnei Politiki Rossii [Archive of Russian Foreign Policy] (Moscow)
f. Agentstvo Ministerstva finansov v Berline
f. II Departamenta
f. Kantseliariia
f. Kitaiskii stol
f. Persidskii stol
f. Politarkhiv
f. Posol'stvo v Vashingtone
f. Posol'stvo v Londone
Hoover Institution Archive (Stanford, California)
Collection: Russian Embassy in Washington ("Perepiska voenno-morskogo agenta" [Correspondence of the Navy's Agent])
LGIA: Leningradskii gosudarstvennyi istoricheskii arkhiv (Leningrad)
f. 1258 Aktsionernoe obshchestvo mashinostroitel'nogo zavoda "L. Nobel'"
Library of Congress (Washington, D.C.)
James R. Garfield papers
Robert Lansing papers
William H. Taft papers
Ministere des affaaires étrangerers. Archive diplomatique (Paris)
Corresponding politique. Russie. Nouvelle serie.
National Archives (Washington, D.C.)
State Department files
Princeton University Library (Princeton, N.J.)
Ivy Lee papers

PRO: Public Record Office (London)
Foreign Office
RAO: Rockefeller Archive Center (Pocantico Hills, Tarrytown, N.Y.)
Rockefeller family
TsGAOR: Tsentral'nyi gosudarstvennyi arkhiv oktiabr'skoi revoliutsii (Moscow)
f. Departament politsii. Osobyi otdel
TsGAVMP: Tsentral'nyi gosudarstvennyi arkhiv voenno-morskogo flota SSSR (Leningrad)
f. 418 Morskoi general'nyi shtab
TsGIA Azerb.SSR: Tsentral'nyi gosudarstvennyi istoricheskii arkhiv Azerbaidzhanskoi SSR (Baku)
f. Kaspiiskoe tovarishchestvo
f. Bakinskoe otdelenie tovarishchestva "Brat'ia Nobel'"
TsGIA Gruz.SSR: Tsentral'nyi gosudarstvennyi istoricheskii arkhiv Gruzinskoi SSR (Tbilisi)
f. 370 Zakavkazskoe aktsiznoe upravlenie
TsGIA SSSR: Tsentral'nyi gosudarstvennyi istoricheskii arkhiv SSSR (Leningrad)
f. 20 Departament torgovli i manufaktur Ministerstva finansov
f. 23 Ministerstvo torgovli i promyshlennosti
f. 37 Gornyi departament (from 1874 to 1905 part of the Ministerstvo gosudarstvennykh imushchestv; after 1905 within the Ministerstvo torgovli i promyshlennosti)
f. 40 Vsepoddanneishie doklady po chasti torgovli i promyshlennosti i torgovye dogovory s inostrannymi gosudarstvennami
f. 560 Obshchaia kantseliariia Ministerstva finansov
f. 574 Departament neokladnykh sborov Ministerstva finansov
f. 588 Peterburgskaia kontora Gosudarstvennogo banka
f. 595 Volzhsko-Kamskii bank
f. 626 Peterburgskii mezhdunarodnyi kommercheskii bank
f. 630 Russko-Aziatskii bank
f. 1263 Komitet ministrov
f. 1450 Obshchestvo Mazut
f. 1458 Tovarishchestvo "Brat'ia Nobel"
f. 1459 Tovarishchestvo Neft'

Published Works Cited

Abels, J. *The Rockefeller Billions* (New York, 1965).
Agahd, E. *Großbanken und Weltmarkt* [Big Banks and the World Market], (Berlin, 1914).
Anan'ich, B. V. "Rossiia i kontsessiia d'Arsy" [Russia and the Concession of d'Arsy], *Istoricheskie zapiski* [Historical Papers], (Moscow, 1960), *66*, pp. 278-290.

Anan'ich, B. V. "Russkoe samoderzhavie i vneshnie zaimy v 1888-1902 gg." [Russian Autocracy and Foreign Loans, 1888-1902], in *Iz istorii imperializma v Rossii* [From the History of Imperialism in Russia] (Moscow-Leningrad, 1959), pp. 183-218.

Anan'ich, B. V. and S. K. Lebedev. "Uchastie bankov v vypuske obligatsii rossiiskikh zheleznodorozhnykh obshchestv (1860-1914 gg.)" [The Participation of Banks in the Issue of Securities for Russian Railway Companies, 1860-1914], in: *Monopolii i ekonomicheskaia politika tsarizma v kontse XIX - nachale XX v.* [Monopolies and the Economic Policies of Tsarism at the End of the Nineteenth and Beginning of the Twentieth Centuries] (Leningrad, 1987), pp. 5-40.

Anan'ich, B. V. and R. Sh. Ganelin. "Opyt kritiki memuarov S. Iu. Vitte (v sviazi s ego publitsisticheskoi deiatel'nost'iu v 1907-1915gg.)" [An Essay in Critical Analysis of S. Iu. Witte's Memoirs (in Connection with His Journalistic Activities in 1907-1915)], in *Voprosy istoriografii i istochnikovedeniia istorii SSSR* [Questions of Historiography and Source Studies in the History of the USSR] (Moscow-Leningrad, 1963), pp. 298-374.

Anan'ich, B. V. "Uchetno-ssudnyi bank Persii v 1894-1907 gg." [The Discount and Loan Bank of Persia, 1894-1907], in *Monopolii i inostrannyi kapital v Rossii* [Monopolies and Foreign Capital in Russia], (Moscow-Leningrad, 1962), pp. 274-314.

Anan'ich, B. V. *Rossiia i mezhdunarodnyi kapital 1897-1914* [Russia and International Capital, 1897-1914] (Leningrad, 1970).

Baumgart, I. von and H. Bennekenstein. "Der Kampf des deutschen Finanzkapitals in den Jahren 1897 bis 1914 für ein Reichspetroleummonopol" [The Struggle of German Finance Capital in the Years from 1897 to 1914 for an Imperial Petroleum Monopoly], in *Jahrbuch für Wirtschaftsgeschichte,* (1980) *2,* pp. 95-120.

Berenger, H. *Le petrole et la France* [Petroleum and France]. (Paris, 1920).

Bovykin, V. I. "Rossiiskaia neft' i Rotshil'dy" [Russian Oil and the Rothschilds], in *Voprosy istorii* [Questions of History], (1978) *4,* pp. 27-41.

Brack, U. *Deutsche Erdölpolitik vor 1914* [German Oil Policy before 1914], (Hamburg, 1977).

Brackel, O. V. and J. Leis. *Der dreißigjährige Petroleumkrieg* [The Thirty-Years War for Oil], (Berlin, 1903).

Brandt, B. .F. *Inostrannye kapitaly. Ikh vliianiia na ekonomicheskoe razvitie strany* [Foreign Capital. Its Influence on the Economic Development of the Country]. Vol. 1. (St. Petersburg, 1898).

British Documents on the Origins of the War, 1898-1914. Edited by G. P. Gooch and H. Temperley, (London, 1926-38).

British Petroleum Statistical Review of World Energy, 1986 (London, 1986).

Campbell, C. S. *Special Busines Interests and the Open Door Policy* (New Haven, 1951).

Challener, R. D. *Admirals, Generals and American Foreign Policy, 1898-1914.* (Princeton, 1973).

Churchill, W. *The World Crisis, 1911-14.* (London, 1923).

Clark, J. S. *The Oil Century*. (Norman, 1958).
D'iakonova, I.A. *Nobelevskaia korporatsiia v Rossii* [The Nobel Corporation in Russia] (Moscow, 1980).
Deterding, H. *An International Oil Man* (London, 1934).
Documents diplomatiques français (1871-1914). [French Diplomatic Documents (1871-1914]. 1 Série. Vol. 13. (Paris, 1953).
Ekonomicheskoe polozhenie Rossii nakanune Velikoi Oktiabr'skoi sotsialisticheskoi revoliutsii. Dokumentv i materialy [The Economic Condition of Russia on the Eve of the Great October Socialist Revolution. Documents and Materials]. (Moscow-Leningrad, 1957).
Engler, R. *The Brotherhood of Oil. Energy Policy and the Public Interest* (Chicago, 1977).
Erusalimskii, A. S. *Vneshniaia politika i diplomatiia germanskogo imperializma v kontse XIX veka* [Foreign Policy and the Diplomacy of German Imperialism at the End of the Nineteenth Century]. (Moscow, 1948).
Eventov, L. *Inostrannyi kapital v neftianoi promyshlennosti Rossii* [Foreign Capital in the Petroleum Industry of Russia]. (Moscow-Leningrad, 1925).
Ferrier, R. *The History of the British Petroleum Co.,* (London, 1982).
Final Report of the Industrial Commission. Vol. 19. (Washington, 1920).
Fink, G. *The Genoa Conference: European Diplomacy, 1921-1922*. (London, 1984).
Fischer, F. *Der Griff nach der Weltmacht* [The Bid for World Power], (Düsseldorf, 1962).
Foley, P. "Petroleum Problems of the World War," *U.S. Naval Institute Proceedings,* (November 1924).
Fursenko, A.A. "Materialy o korruptsii tsarskoi biurokratii" [Materials on the Corruption of the Tsarist Bureaucracy], *Issledovaniia po otechestvennomu istochnikovedeniiu* [Research on Source Analysis for (Russian) National History], (Moscow-Leningrad, 1964), pp. 149-156.
Fursenko, A. A. "Mozhno li schitat' kompaniiu Nobelia russkim kontsernom?" [Can the Nobel Company Be Considered a Russian Enterprise?], *Issledovaniia po sotsial'no-politicheskoi istorii Rossisi* [Studies on the Social and Political History of Russia], (Leningrad, 1971), pp. 149-156.
Fursenko, A. A. *Neftianye tresty i mirovaia politika (1880-e-gody—1918 g.)* [Oil Trusts and World Politics (1880-e—1918)] (Leningrad, 1965).
Fursenko, A. A. "Pervyi neftianoi eksportnyi sindikat v Rossii" [The First Oil Exporting Syndicate in Russia] *Monopolii i inostrannyi kapital v Rossii* [(Monopolies and Foreign Capital in Russia], (Moscow-Leningrad, 1962), pp. 4-58.
Gardner, L. C. *Wilson and Revolutionaires, 1913-1921*. (Washington, D.C., 1976).
Gefter, M. Ia., L. E. Shepelev and A. M. Solov'eva. "O proniknovenii angliiskogo kapitala v neftianuiu promyshlennost' Rossii (1989-1902 gg.)" [On the Penetration of English Capital into the Petroelum Industry of Russia (1898-1902)], *Istoricheskii arkhiv* [Historical Archive], (1960), *6,* pp. 76-104.
Gerretson, F. C. *History of the Royal Dutch*. 4 vols. (Leiden, 1958).
Gibb, G.S. and F. M. Knowlton. *The Resurgent Years 1911-1927*. (New York, 1956).

Gindin, I. F. "Ob osnovakh ekonomicheskoi politiki tsarskogo pravitel'stva v konste XIX—nachale XX v." [On the Foundations of the Economic Policy of the Tsarist Government at the End of the Nineteenth and Beginning of the Twentieth Century], *Materialy po istorii SSSR* [Materials on the History of the USSR] (Moscow, 1959), *6*, pp. 159-222.

Gofman, K. *Neftianaia politika i anglo-saksonskii imperializm* [Petroleum Policy and Anglo-Saxon Imperialism] (Leningrad, 1930).

Graebner, N. A., (ed.), *An Uncertain Tradition. American Secretaries of State in the Twentieth Century* (New York, 1961).

Haase, F. *Die Erdöl-Interessen der Deutschen Bank und der Direktion der Disconto-Gesellschaft in Rumänien* [The Petroleum Interests of the Deutsche Bank and the Directors of the Disconto-Gesellschaft in Rumania], (Berlin, 1922).

Hagelin, K.W. *Moi trudovoi put'* [My Professional Experiences]. (New York, 1945).

Hallgarten, G. W. F. *Imperialismus vor 1914. Die soziologischen Grundlagen der Außenpolitik europäischen Großmächte vor dem Ersten Weltkrieg* [Imperialism before 1914. The Sociological Foundations of the Foreign Policy of the Great Powers in Europe Prior to the First World War]. Vol. 1. (Munich, 1963).

Hardinge, A. *A Diplomatist in the East.* (London, 1928).

Hansard's Parliamentary Debates, (London, 1913-1914).

Hawke, D.F. *John D. The Founding Father of the Rockefelers.* (New York, 1980).

Hewins, R. *Mr. Five Per Cent. The Story of Calouste Goulbenkian,* (New York, 1958).

Hidy, R. and M. Hidy. *Pioneering in Big Business,* (New York, 1955).

The Intimate Papers of Colonel House. Arranged as Narrative by Charles Seymour. 4 vols. (Boston, New York, 1926).

Istoriia Azerbaidzhana [The History of Azerbaijan]. Vol. 2. (Baku, 1960).

Jones, G. *The State and the Emergence of the British Oil Industry.* (London, 1981).

Jones, G. and C. Trebilcock. "Russian Industry and British Business, 1910-1930: Oil and Armaments," *Journal of European Economic History, 11* (1982).

Katz, F. *Deutschland, Diaz and die mexikanische Revolution* [Germany, Diaz and the Mexican Revolution] (Berlin, 1964).

Katz, F. *The Secret War in Mexico. Europe, the United States and the American Revolution* (Chicago, 1981).

Kent, M. *Oil and Empire. British Policy and Mesopotamian Oil, 1900-1920.* (London, 1976).

L'vov, L. [Kliachko, L.M.] *Za kulisami starogo rezhima. Vospominaniia zhurnalista* [Behind the Scenes under the Old Regime. Memoirs of a Journalist]. (Leningrad, 1926).

Lebedev, S. K. "Peterburgskii mezhdunarodnyi kommercheskii bank v konsortsiumakh po vypusku chastnykh zheleznodorozhnykh zaimov 1880-kh—nachala 1890-kh gg." [The St. Petersburg Internaitonal Commercial Bank in Corsortia for the Issue of Private Railway Loans (1880s-Early 1890s)], *Monopolii i ekonomicheskaia politika tsarizma v knotse XIX—*

nachale XX v. [Monopolies and the Economic Policy of Tsarism at the End of the Nineteenth and Beginning of the Twentieth Centuries], (Leningrad, 1987), pp. 41-65.

Lenin, V. I. *Polnoe sobranie sochinenii* [Complete Collection of Works]. 55 vols. (Moscow, 1958-1965).

Longhurst, H. *Adventure in Oil. The Story of British Petroleum* (London, 1959).

McKay, J. P. "Enterpreneurship and the Emergence of the Russian Petroleum Industry, 1813-1883," *Research in Economic History,* (1982), *8,* pp. 47-91.

Mai, J. *Das deutsche Kapital in Rußland, 1850-1895* [German Capital in Russia, 1850-1895]. (Berlin, 1870).

Marder, A. J. *From the Drednaught to Scapa Flow. The Royal Navy in the Fisher Era, 1904-1919.* vol. 1. (London, 1961).

Monopolisticheskii kapital v neftianoi promyshlennosti Rossii. 1883-1914. Dokumenty i materialy [Monopolistic Capital in the Petroleum Industry of Russia, 1883-1914. Documents and Materials], (Moscow-Leningrad, 1961).

Moody, J. *The Truth about Trusts* (New York, 1904).

Moore, A. L. *John D. Archbold and the Early Development of Standard Oil* (New York, n.d.).

Nardova, V. A. *Nachalo monopolizatsii neftianoi promyshlennosti Rossii. 1880-1890-e-gody* [The Beginnig of the Monopolization of the Petroleum Industry in Russia. The 1880s and 1890s] (Leningrad, 1974).

Nevins, A. *John D. Rockefeller. The Heroic Age of American Enterprise.* vol. 1. (New York, 1940).

Nevins, A. *A Study in Power. John D. Rockefeller, Industrialist and Philanthropist.* 2 vols. (New York, 1953).

Niring, S. [Nearing, Scott]. "Vneshniaia politika Soedinennykh Shtatov. Doklad v MGU 23 noiabria 1927 b." [The Foreign Policy of the United States. Paper Presented at Moscow State University on 23 November 1927], *Mirovoe khoziaistvo i mirovaia politika* [The World Economy and World Politics], (1928), *1,* pp. 54-60.

O'Connor, H. *The World Crisis in Oil* (London, 1963).

Obzor bakinskoi neftianoi promyshlennosti za 1914 g. [An Overview of the Baku Oil Industry for 1914] (Baku, 1915).

Obzor bakinskoi neftianoi promyshlennosti za 1901 g. [An Overview of the Baku Oil Industry for 1901]. (Baku, 1902).

The Oil War in Mexico. History of a Fortune-Wrecking Fight Between Colossal Interests Headed by Mr. H. C. Pierce and Sir W. Pearson. (New York, 1910).

Ol', P. V. *Inostrannye kapitaly v narodnom khoziaistve dovoennoi Rossii* [Foreign Capital in the National Economy of Prewar Russia]. (Leningrad, 1925).

Olstead, A. L. and P. Rhode. "The U.S. Energy Crisis of 1920 and the Search for new Oil Supplies," in *Oil in the World Economy. The Ninth International Economic History Congress,* edited by R.W. Ferrier and A.A. Fursenko (Bern, 1986), pp. 22-23.

Pershke, S. and I. *Russkaia neftianaia promyshlennosti', ee razvitie i sovremennoe polozhenie v statisticheskikh dannykh* [The Russian Petroelum Industry, Its Development and Current Condition in Statistical Data]. (Tiflis, 1913).

Polovtsev, A. A. "Dnevnik" [Diary], *Krasnyi arkhiv* [Red Archive], (1931), *4*, pp. 110-132.

Prager, M. "Die amerikanische Gefahr" [The American Danger], in *Volkswirtschaftliche Zeitfragen* [Economic Topics of Today], Jahrgang XXIV (Berlin, 1902).

Primakov, A. E. *Persidskii zaliv: neft' i politika.* [The Persian Gulf: Oil and Politics] (Moscow, 1983).

Revoliutsiia 1905-1907 gg. v Rossii. Dokumenty i materialy [The Revolution of 1905-1907 in Russia. Documents and Materials]. Pt. 2. (Moscow, 1955).

Romanov, B. A. *Ocherki diplomaticheskoi istorii russko-iaponskoi voiny. 1895-1907gg.* [Sketches of the Diplomatic History of the Russo-Japanese War. 1895-1907]. (Moscow-Leningrad, 1956).

Schwadran, B. *The Middle East, Oil and the Great Powers.* (New York, 1955).

Seltzer, L. H. *A Financial History of the American Automobile Industry.* (New York, 1928).

Shannon, F. A. *America's Economic Growth.* (New York, 1947).

Shishkin, V. A. "Soviet Oil Exports between Two World Wars," in *Oil in the World Economy. The Ninth International Economic History Congress,* edited by R. W. Ferrier and A. A. Fursenko (Bern, 1986), pp. 14-22.

Solov'ev, Iu. B. "Peterburgskii mezhdunarodnyi bank i frantsuzskii finansovyi kapital v gody pervogo promyshlennogo pod"ema v Rossii" [The Petersburg International Bank and French Financial Capital in the Years of the First Industrial Boom in Russia], *Monopolii i inostrannyi kapital v Rossii* [Monopolies and Foreign Capital in Rusia] (Moscow-Leningrad, 1962), (1962), pp. 377-407.

Solov'ev, Iu. B. "Protivorechiia v praviashchem lagere Rossii po voprosu ob inostrannykh kapitalakh v gody pervogo promyshlennogo pod"ema" [Contradictions in Russia's Ruling Elites on the Question of Foreign Capital (during the years of the First Industrial Boom)], *Is istorii imperializma v Rossii* [From the History of Imperialism in Russia] (Moscow-Leningrad, 1959), pp. 371-388.

Solov'ev, Iu. B. "Franko-russkii soiuz v ego finansovom aspekte (1895-1900 gg.)" [The Franco-Russian Alliance in Its Financial Dimension (1895-1900)], *Frantsuzskii ezhegodnik. 1961 g.* [The French Yearbook for 1961] (Moscow, 1962), pp. 163-206.

Tolf, R. W. *The Russian Rockefellers* (Stanford, 1976).

Tugendhat, C. *Oil. The Biggest Business* (New York, 1968).

U.S. Commercial Relations. (Washington, D.C., 1894-1905).

U.S. Consular Reports. (Washington, D.C., 1894-1905).

U.S. Mineral Resources. (Washington, D.C., 1894-1905).

United Nations Statistical Yearbook: 1980. (New York).

Vagts, A. *Mexico, Europa and America unter besonderer Berücksichtigung der Petroleumpolitik* [Mexico, Europe and America (with Special Attention to Petroleum Policies)] (Berlin, 1928).

Vagts, A. *Deutschland und die Vereinigten Staaten in der Weltpolitik* [Germany and the United States in World Politics]. Vol. 1-2. (New York, London, 1935).

Vilenkin, G. *Finansovyi i ekonomicheskii stroi sovremennoi Anglii* [The Financial and Economic Order of Contemporary England] (St. Petersburg, 1902).

Vol'skii, V. *Latinskaia Amerika, neft' i nezavisimost'* [Latin America, Oil and Independence] (Moscow, 1964).

Volobuev, P. V. "Iz istorii monopolizatsii neftianoi promyshlennosti dorevoliutsionnoi Rossii (1903-1914 gg.)" [From the History fo the Monopolization of the Petroelum Industry in Prerevolutionary Russia (1903-1914)], *Istoricheskie zapiski* [Historical Papers], (1955), *55,* pp. 80-111.

Von Laue, T. "A Secret Memorandum of Sergei Witte on the Industrialization of Imperial Russia," *Journal of Modern Hisory,* (1954), *36,* pp. 67-74.

Von Laue, T. *Sergei Witte and the Industrialization of Russia.* (New York, 1963).

Wigham, H. J. *The Persian Problem,* (London, 1903).

Wilkins, M. *The Emergence of the Multinational Enterprise.* (Cambridge, 1970).

Williamson, H. F. and Daum, A. R. *The American Petroleum Industry.* Vol. 1: *The Age of Illumination, 1859-1899* (Evanston, 1959) Vol. 2: *The American Petroleum Industry. The Age of Energy, 1899-1959* (Evanston, 1963).

Witte, S. Iu. *Vospominaniia* [Memoris]. 3 vols. (Moscow, 1960).

Zarnitskii, S. V. and L. I. Trofimova. *Sovetskoi strany diplomat* [A Diplomat of the Soviet Union]. (Moscow, 1968).

Index